In memory of two giants of mathematics and natural science,
Henri Poincare and Aleksandr Lyapunov

Aleksandr Mikhailovich Lyapunov
(06/06/1857 – 11/03/1918)

Jules Henri Poincaré
(04/29/1854 – 08/17/1912)

Aleksandr Mikhailovich Lyapunov was a Russian mathematician, mechanician and physicist. He is a founder of the Mathematical Theory of Stability and the author of the Central Limit Theorem in Probability Theory. Lyapunov's research has been having a tremendous impact on the development of modern science and technology since the birth of the era of aviation.

Jules Henri Poincaré was a French mathematician, physicist and engineer. He is considered as the last Universalist in mathematics and the creator of the new branch of mathematics, Topology, as well as Qualitative Theory of Differential Equations. Poincaré's scientific and philosophical views have been astoundingly determining the evolution of mathematics and physics for over a hundred years.

The theoretical works of A. M. Lyapunov and J. H. Poincaré have been tacitly keeping on making a prodigious contribution to shaping the future of our world.

CONTENTS

PREFACE

My passion for Lyapunov functions was born in the remote 1986 when as a freshman of PhD program of Kiev Engineering College of Air Force in its science and technology library I came across the monograph concerned with the Lyapunov's second method of the investigation of stability written by my thesis advisor Professor Nikolay Sergeevich Sivov. After reading and falling in love with it I told him that I had intentions to use the method in my research. I remember his answer very well, "Lyapunov functions are like a miracle from the Heavens. If they come to scientists or engineers then they can do the impossible. Unfortunately, this miracle happens to us very rare just because we are not ready for it yet." The words did not discourage or disappoint me. They challenged me and made my passion secret. I kept them in the back of my mind for almost two decades. Working under the supervision of Professor N. S. Sivov was a great joy because of his wonderful personal trait to instill us, young researchers, with his overwhelming enthusiasm, remarkable curiosity and strong resolution to handle extremely difficult or even insurmountable problems without any fear. "Look for something unusual, strange in your problem. Peculiarities usually are the keys to them." was his scientific motto. Somewhere around 2006, eleven years after laying down the foundations of the theory of nonlinear non-autonomous dynamic systems of generic configuration, I realized that this result can help us become ready for the miracle of Lyapunov functions. The term of "generic configuration" means that the extended phase spaces of the dynamic systems have nor equilibrium points neither all other kinds of degeneracies or their motions take place in the domains located so "far away" from them that the method of linearization does not work. My train of thoughts approximately was as follows. Let us recall the original work of A. M. Lyapunov. In modern terms he considers a given dynamic system, autonomous one $\frac{dx}{dt} = f(x)$ or non-autonomous one $\frac{dx}{dt} = f(t,x)$, $\forall t \in [t_0;+\infty[$ and with a particular solution $x = x(t;\hat{x}_0)$ being under investigation for stability, where $x = \hat{x}_0 = x(t_0;\hat{x}_0)$ is the initial point and $\dim x = n$. Utilizing the transformation $\vartheta : (t, x = z + x(t;\hat{x}_0)) \to (t,z)$, where $\dim z = \dim x = n$, A. M. Lyapunov rectifies the solution $x = x(t;\hat{x}_0)$ and brings it to the origin of the orthogonal

Cartesian coordinate system, where the system is described by ordinary differential equations. He calls the transformed and in any case non-autonomous system $\frac{dz}{dt} = \tilde{f}(t,z)$ with $z \equiv 0$ as a particular solution under investigation the perturbed one and continues with analyzing just it. While the initial step was the impeccable stroke of genius, nevertheless the above-mentioned transformation is strictly local. It does not take into consideration the properties of respectively $(n\text{-}1)$ and (n) one-codimensional invariant manifolds or invariant hypersurfaces that form the solution $x = x(t;\hat{x}_0)$. Thus, we terribly needed some global diffeomorphism that takes account of and conserve all these properties during transformation. But how to find it? Strange as it may seem but the answer was floating on the surface for 122 years. If the Lyapunov's initial step was perfect then what we just needed to do was to advance consistently in this direction rectifying or flattening the invariant hypersurfaces exactly as we did with the integral curve $x_t(\hat{x}_0) = (t, x(t;\hat{x}_0))$. This diffeomorphism converts the invariant hypersurfaces to invariant hyperplanes that are also manifolds but with the simplest possible internal geometry. The only thing remained to figure out was the procedure of flattening hypersurfaces that conserves the properties. Using modern mathematical terminology, it was necessary to find out the diffeomorphism with certain topological and algebraic invariants. To our luck the theory of nonlinear non-autonomous dynamic systems of generic configuration is based on the study of what we were in desperate search, namely maps between the one-codimensional invariant manifolds and how the former influence their topological, geometrical and algebraic properties. Thanks to this fact the procedure has been successfully devised and named the cascade of sequential flattening diffeomorphisms. The Figures A, B, C given below illustrate geometrically in three-dimensional extended phase space with two phase variables and the time t the leap forwards from the absolutely local Lyapunov's transformation that rectifies only the integral curve to the aforementioned procedure, which not only rectifies the one but also flattens the one-codimensional invariant manifolds forming it. The Figure A shows the typical geometrical configuration of two one-codimensional invariant manifolds or invariant surfaces M_{x_1} and M_{x_2} in the three-dimensional space, which intersection is the integral curve x_t of a given original three-dimensional dynamic system. The Figure B demonstrates what has happened to the configuration of the invariant surfaces and the integral curve denoted by M_{z_1}, M_{z_2}, z_t

respectively in the new coordinate system after applying the Lyapunov's transformation ϑ to M_{x_1}, M_{x_2} and $x_t = \bigcap_{i=1}^{2} M_{x_i}$. Here the integral curve turns in a straight line or more precisely in the ray $z_t = (t, z_1 = 0, z_2 = 0)\ \forall t \geq 0$ emitted from the origin of the orthogonal Cartesian coordinate system and moving along the axis t towards $+\infty$. However the invariant surfaces remain curved. The Figure C illustrates the result of the application of the cascade of sequential flattening diffeomorphisms θ. This time the integral curve y_t also is the ray as in the previous case but the invariant surfaces convert into the planes $\{y_1 = 0\}$ and $\{y_2 = 0\}$. This is a pivotal point of the research because under the action of the cascade of sequential flattening diffeomorphisms θ, which can be presented as the canonizing diffeomorphism $\overline{\varphi}$, the original system assumes a canonical form. The last point allows to introduce the canonical forms of Lyapunov functions and obviating the need for racking the brains how to find these unfathomable ones.

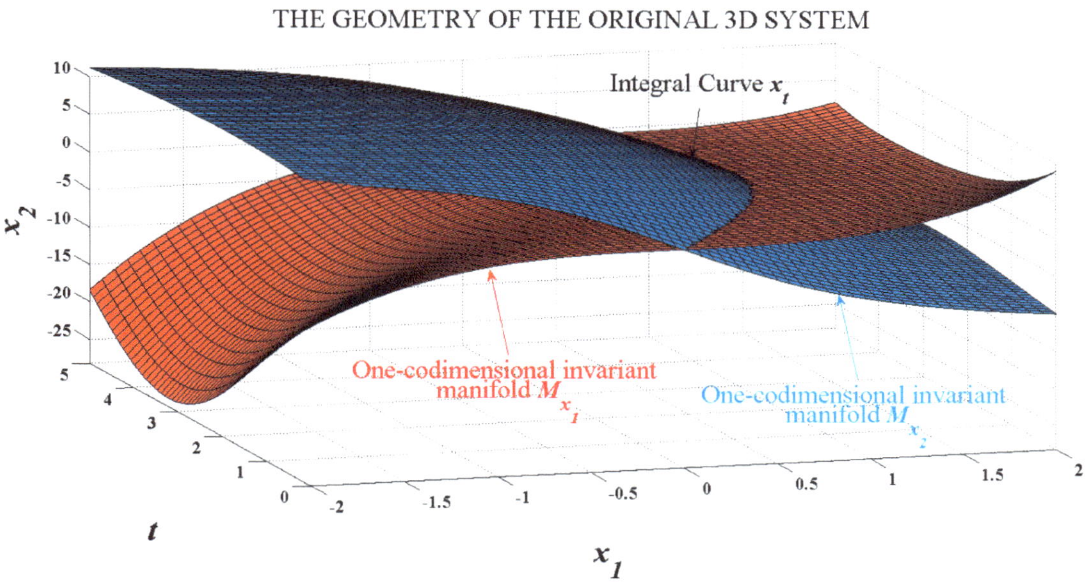

FIG. A

THE GEOMETRY OF THE 3D ORIGINAL SYSTEM AFTER THE LYAPUNOV'S TRANSFORMATION
$\vartheta: (t, x_1, x_2) \to (t, z_1, z_2)$

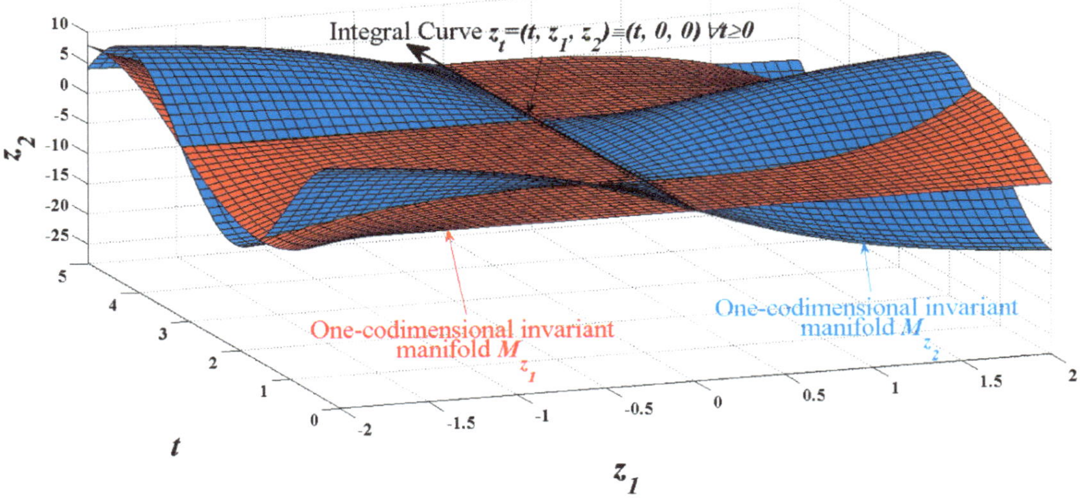

FIG. B

THE GEOMETRY OF THE 3D ORIGINAL SYSTEM TRANSFORMED BY THE CASCADE OF SEQUENTIAL FLATTENING DIFFEOMORPHISMS $\theta: (t, x_1, x_2) \to (t, y_1, y_2)$

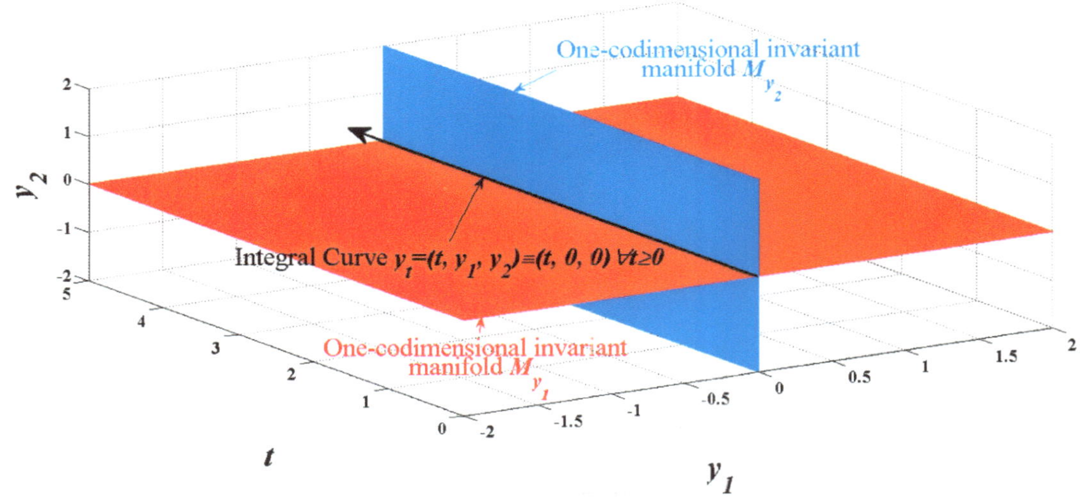

FIG. C

The idea just described above is that of several underlying ones, which compose the core concept used in the research that has opened the door to the breakthrough in making *le miracle quotidian* out of *the miracle of Lyapunov functions from the Above*. It took more eight years to develop the theory and put it on paper.

The engineering schools of Kiev Engineering College of Air Force, Kiev Polytechnic Institute, Institut Aéronautique et Spatial, École Nationale Supérieure d'Ingénieurs de Constructions Aéronautiques and Research Center of General Staff together with the mathematical schools of Institute of Mathematics and Institute of Cybernetics of National Academy of Science was the making of me as a mathematician and engineer. During the formative years spent at these highly respected educational and research institutions I met many wonderful scientists and educators who had a significant influence on me. I would like to pay tribute to them here by mentioning their names: N. S. Sivov, A. I. Lysenko, V. P. Prokofiev, A. E. Aslanyan, V. V. Sharko, Yu. A. Mitropolskiy, V. M. Kuntsevich, A. I. Kukhtenko, B. N. Pshenichny, O. S. Uruskiy, M. A. Pavlovskiy, E. S. Scherbina, Joël Bordeneuve-Guibé, Jean Louis Abatut, Didier Feriol, Michel Chauvin, Alain Riesen.

Lastly, I wish to thank my wife, Olga, who has been staunchly standing by me for so many years.

Myroslav K. Sparavalo
New York City
April 12, 2016

THE CORE CONCEPT AND THE OBJECTIVES

There are two hurdles on the way to a general method of construction of Lyapunov functions that must be cleared. The number one relates to the first step in this direction when we face the $n+1$-dimensional dynamic systems governed by the non-autonomous nonlinear ordinary differential equations (ODEs) of the general form as follows

$$\frac{dx}{dt} = f(t,x).$$

What should we do in the very beginning? What mathematical manipulations should we make on the initial step? What information do we need to start moving? There had been no answers to these questions before my paper "The Lyapunov Concept of Stability from the Standpoint of Poincaré Approach: General Procedure of Utilization of Lyapunov Functions for Non-Linear Non-Autonomous Parametric Differential Inclusions" was published on http://arxiv.org/abs/1403.5761. In some specific cases analytical expressions for the conservation laws, the potential and kinetic energy, Hamiltonians can help with finding Lyapunov functions. But in fact all these expressions are first integrals or their functional combinations. First who explicitly proposed to use the bundles of first integrals for constructing Lyapunov functions was N. G. Chetaev. His guess being rather empirical than deductive came out in 1946 when the first edition of the acclaimed book "Stability of Motion" was published. To be honest completely we have to admit that it was A. M. Lyapunov who was first to have been reflecting on the utilization of first integrals for constructing his functions. The example with the system governed by the differential equations

$$\left\{\frac{dx_i}{dt} = \frac{\partial V(x_1,\dots,x_n)}{\partial x_i}, i = (1,\dots,n)\right\}$$

in his famous thesis "General Problem of the Stability of Motion " reveals the thoughts and intentions of the Mastermind. However it was just a hint or a puff of the idea only. That is why its priority remains on N. G. Chetaev.

There have been numerous attempts after him to develop his vision of the Lyapunov's second method. Unfortunately, in the general case, as the system given above, all these endeavors have not been crowned with success. All the cunning tricks have turned out to be useless. Moreover, it was impossible to grope around for something even very-very distantly similar to a foothold. Moving closer to 21st century the mathematical theory of stability was insistently calling for a new idea that could bind together first integrals and Lyapunov functions.

My work has showed what we need else to create a general method of construction of Lyapunov functions. This is the full set of the first integrals of the system or at least their corresponding sections forming the integral curve, which we want to probe on stability. The system has (n) first integrals depending on t. Their sections are n-dimensional invariant manifolds or invariant hypersurfaces. And here the second hurdle is coming out. There is no general method for integrating ODEs and therefore there is no general method of finding first integrals. However, if we limit our appetite to the sections of the first integrals, then the problem becomes much-much simpler. First, the sections or more exactly only their pieces in the neighborhood of the integral curve can be found or constructed by means of numerical methods. In any case, on the modern level of development of mathematics we cannot proceed without computer modeling and simulation. Second, in my earlier works I proposed the concepts of proper manifold and its influence set constructed with help of the concept of equidifferential manifold. In practice we do not need the exact expressions for invariant hypersurfaces. We need only some admissible approximations of them. It is these approximations, for constructing which the above-mentioned geometric-topological concepts were designed.

Thus, we can state that a general method of construction of Lyapunov functions has shifted from the area called *"I have no idea how even to start handling this problem!"* to the area of encompassable problems within our contemporary mathematical and technological abilities. The first hurdle has been passed successfully. For the second one we have certain strategy to overcome.

Now we will pay our attention to how the first obstacle has been cleared. A frontal attack on the problem of the creation of a general method of construction of Lyapunov functions is a Sisyphus labor. It is supposed to seek the functions manipulating the right-hand sides written in the most general form of $f(t,x)$. All the previous experience of a number of the generations of researchers proved the futility of such efforts. *Our core concept reverses the course of the process of solving the problem from the initial point to the final one.* The commonly accepted formulation of the problem states that there is a given initial system $\frac{dx}{dt} = f(t,x)$ and it is necessary to find a Lyapunov function for it. Our thought assumes that we already have some standard Lyapunov functions. Now our objective is to find a special representation of the initial system so that we could apply the theorems of the Lyapunov's second method to it as the first step. This special representation should be a result of the application of some special transformation to the initial system. The second step exploits the fact that the initial system and its special representation are tied together by the special transformation to judge the stability of the former by using our evaluation of the stability of the latter made at the first stage. In the monograph the words of "standard" and "special" are replaced with the words of "canonical" and "canonizing". Thus, we have the canonical Lyapunov functions and the canonizing transformation that establishes certain relation between the initial system and its canonical form. The ultimate point in all these manipulations is to investigate the system of canonical form by means of the canonical Lyapunov functions and due to the established relation to be able to carry the end result of the investigation about its stability over into the initial system. Actually, this relation should be at least the inverse equivalency of the presence of the property of stability in two forms (initial and canonical) of our system to an equal extent. Specifically, it means that if for the canonical form of our system there is certain canonical Lyapunov function ensuring its asymptotical stability then the system of initial form or in other words the initial one is also asymptotically stable.

But what if the problem of finding the canonical forms of Lyapunov functions is insuperable just like the one of first integrals is? Luckily, we have a clue. For a one-dimensional system the simplest auxiliary Lyapunov function can take the form of $(x_1)^2$. Extending this idea to the $n+1$-dimensional case we can suppose that here the simplest auxiliary Lyapunov function takes the form of a sum of all $(x_i)^2$, namely $\sum_{i=1}^{n}(x_i)^2$. In Chapter 3 we will show that if an auxiliary Lyapunov function $W(x)$ is found then the Lyapunov function $V(t,x)$ can be easily constructed on its base.

Now what should we do with the initial system to obtain its canonical form? As we said before any integral curve is an intersection of n-dimensional manifolds being the sections of the corresponding first integrals. There is the only property they have that hinders us from advancing in this direction. They are not flat. *Why is it so important to transform them into hyperplanes?* The answer comes from the expression for a full derivative of the auxiliary Lyapunov function $W = \sum_{i=1}^{n}(x_i)^2$ if the n-dimensional invariant manifolds are hyperplanes.

We have $\frac{dW}{dt} = 2\cdot\sum_{i=1}^{n} x_i \cdot f_i(t,x)$, where $f_i(t,x) = 0$, if $x_i = 0$. It is clear that for the asymptotical stability of such a system it is sufficient that the components x_i and $f_i(t,x)$ of each term $x_i \cdot f_i(t,x)$ of the sum have the opposite signs. That is why we need the procedure of sequential flattening of invariant hypersurfaces. We transform them into invariant hyperplanes one-by-one in a rigor order. It is necessary to keep all these things in mind when reading the monograph. The only difference exists between this clarification of the main idea and the monograph. In the latter the right-hand sides of ordinary differential equations depend on parameters. But this difference changes nothing in the explanation because it is given in case the vector of parameters has taken a specific value and this is the first part of the research. In the second part one considers how the perturbations of the vector of parameters affect the property of stability transferring through the geometric-topological pentad illustrated graphically below.

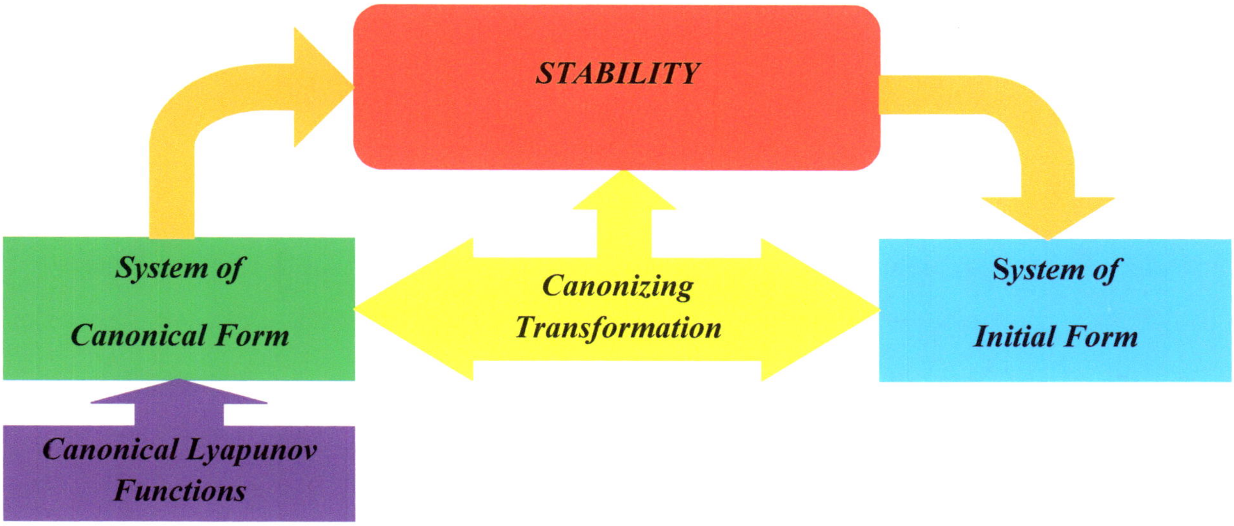

The first objective of the monograph is to present a general method of constructing Lyapunov functions for non-linear non-autonomous differential inclusions described by ordinary differential equations with parameters. We attain the goal via the following ideas synthesizing the core concept:

1. Three-point Poincare's strategy of the study of differential equations and manifolds.
2. The geometric-topological structure of the non-linear non-autonomous parametric differential inclusions in the framework of hierarchical fiber bundles.
3. The canonical Lyapunov functions or the ones of standard forms.
4. The special canonizing transformation of the differential inclusions that presents them in special canonical form, for which the standard forms of Lyapunov functions exist.
5. The conditions of the equivalency establishing the relation between the local asymptotical stability of two corresponding particular integral curves of a given differential inclusion written in the initial and canonical forms.

We will investigate the global asymptotical stability of the entire free dynamical systems as some restrictions of a given parametric differential inclusion and the whole latter one per se in terms of the classificational stability of the typical fiber of the metabundle. At the end of the monograph the prospects of development and modifications of the Lyapunov's second method in the light of the discovery of its new features will be discussed.

The second objective is defined by the author's intentions to open up new horizons in order to unlock the almost omnificent potential of Poincaré's approach to the method of Lyapunov functions for the problems of Mathematical Modeling and Simulation, Non-linear Dynamics and Control Design. Its remarkable efficiency has been revealed in the synthesis of the asymptotically stable in the large and wide-sense robust control of a van der Pol forced dynamic system. The introduced concept of the wide-sense robustness takes account of the full or partial loss of control resources making the control missions still possible after major damages have been done to the dynamic systems with control affected by the unfavorable changes of environment and hostile external disturbances. It is difficult to overestimate the significance of the presence of this intrinsic system characteristic under such adverse conditions of functioning. The second part of the book takes the readers down-to-earth and convincingly proves the amazingly great power of Poincaré's approach to the method of Lyapunov functions at designing control systems satisfying the very high standards of operation performance requirements.

"… the mystery that has been kept hidden for ages and generations, but is now disclosed …",

Colossians 1:26

1. INTRODUCTION

Stability together with controllability on many occasions composes the core of research interests in the modern science and technologies where the mathematical models of physical phenomena are used to break new ground. Today even in medicine the process of successful healing and recovering is described by mathematical models of living organisms able despite the influence of harmful external perturbations to return to their initial healthy state. Sometimes to achieve the complete curing it is necessary to solve the inverse problem of stability namely, to make damaged genes lose the stability of the transmission, regeneration and restoration of their "nefarious properties". Unfortunately high dimensionalities, nonlinearity, parametric and structural uncertainty of mathematical models create a lot of troubles on the way to solving problems of stability. In fact by now there is only one method that allows to tackle the problems under such tough conditions in the most general case. This is the Lyapunov's second method or Lyapunov functions. But since the time, that is about a hundred years ago, when A. M. Lyapunov introduced the special functions later named after him in the theory of stability the problem of finding or constructing them in the most general case has seemed desperately insurmountable. There is no wonder because one does not exist any general method of finding the analytical solutions of the systems of nonlinear non-autonomous differential equations. This fact has endowed Lyapunov functions with a sort of mystery. It has been unclear how Lyapunov functions relate to the first integrals of the systems of differential equations if they do in general. They have always produced the impression of some kind of artificiality imposed on the systems. It has been considered that even if having mostly extraneous rather than the innate relation to the systems, Lyapunov functions may be interjected with or interpreted as part of their intrinsic essence. And mathematicians and engineers have had to put up with this uncertainty for almost a century. However it transpires that the Henri Poincare's strategy of the investigation of differential equations and manifolds allows us to handle successfully even such hopeless problems.

It would be expedient to remind his approach.

1. Henri Poincaré preferred the non-parametric form of the manifold representation to parametric one namely, a m-dimensional submanifold is defined by a system of l implicit

functional equations, each of which depends of n variables and describes the corresponding one-codimensional or $(n - 1)$-dimensional manifold. The following relation holds true in this case: $m = n - l$. Thus the m-dimensional submanifold can be represented by the intersection of l of the one-codimensional manifolds.

2. He suggested to carry out the investigation of m-dimensional submanifold beginning with the study of each of l of the one-codimensional manifolds, which, intersecting, form it. Why? Since as a part of the latter ones the former inherits their properties in certain way.
3. The right-hand sides of the differential equations completely define the qualitative and quantitative behavior the dynamical systems they describe.

Now let us formulate the objectives of the proposed research.

First, we have intentions to establish the relation between Lyapunov functions and the intrinsic properties of the systems of differential equations carried by their right-hand sides and/or first integrals. Second, we would like to develop the general method of constructing Lyapunov functions regardless of the dimensionality, all kinds of nonlinearity and autonomousness of the dynamic systems. In order to expand the range of the generality of the investigated systems of ordinary differential equations not containing controlling functions in their right-hand sides to the possible maximum we will consider parametric differential inclusions.

Our plan of actions is the following one.

First, we will give certain geometric-topological representation to the nonlinear dynamics of parametric differential inclusions, where the center of investigative interest will be integral curves and invariant manifolds of various dimensions. This goal will be attained in the framework of fiber bundles and foliations. Second, we will find a special transformation of the initial parametric differential inclusions that allows us to render it certain "convenient and universal" form to handle the problem of stability further. The "convenience and universality" should be understood in the following sense. Figuratively, the stability can be explained as the ability of systems to produce the limited and even tending to zero with time system reaction to the various input disturbances ranging from small to large magnitudes. It is the wide variety of input disturbances together with nonlinearity are the principal real plagues in the problem of finding and constructing Lyapunov functions. The special "convenient" transformation is meant to "tuck" all these plagues into the right-hand sides of the transformed parametric differential inclusions. Third, we will establish certain correspondence of stability between the initial and transformed parametric differential inclusions in the following sense: if the latter ones behave as asymptotically stable then the former ones also do. Then we will find the Lyapunov functions of so-called canonical form that are universal for all the class and types of the transformed parametric differential inclusions. Forth, the last thing we plan to do is to obtain the conditions for the local asymptotical stability of integral curves, the global asymptotical stability for dynamic systems and parametric differential inclusions, for which the dynamic systems are some restrictions when the vector of parameters of the inclusion assumes certain values.

2. BASIC GEOMETRIC-TOPOLOGICAL STRUCTURES OF PARAMETRIC DIFFERENTIAL INCLUSIONS IN THE FRAMEWORK OF FIBER BUNDLES

Consider a differential inclusion as follows

$$\frac{dx}{dt} \in \bigcup_{\xi \in \Xi} f(t,x;\xi), \tag{1}$$

where $t \in [t_0;+\infty[= T \subset R_t^1$ is time and for simplicity we will consider the ray T unchangeable; $x = (x_1,...x_n) \in R_x^n$ is a phase vector; R_t^1 and R_x^n are Euclidean phase spaces with Cartesian rectangular coordinate systems; $\xi = (\xi_1,...,\xi_m) \in \Xi \subseteq R_\xi^m$ is a vector of parameters, which values belong to some compact open parameter manifold Ξ with $\dim \Xi = m$, $m \geq n$; $(t,x;\xi) \in T \times R_x^n \times \Xi$; $f(t,x;\xi) = (f_1(t,x;\xi),..., f_n(t,x;\xi)) \in C^r$ is a vector-function of indicatrix field, $r \geq 1$.

Let us make some introductory conventions and designations. First of all, basic geometric-topological structures of parametric differential inclusions will be examined as it was done in [1]. Denote $\xi_0 = (\xi_{1,0},...,\xi_{m,0}) \in \Xi$ some given nominal or distinguished point of the vector of parameters ξ. The necessity of the introduction of the point can be substantiated by considering its neighborhood $\Xi_\xi \subseteq \Xi$, when we tackle problems of local stability. Make the convention that the mathematical symbol "hat" or "^" denotes some specific value of any variable or vector considered in the book. For example, $\hat{\xi}_0$ designates a specific value of the distinguished point $\xi_0 \in \Xi$ and $\hat{\xi} \neq \hat{\xi}_0$ is any concrete value of the vector ξ different from the distinguished $\hat{\xi}_0$; $\hat{x}_0 = (\hat{x}_{1,0},...\hat{x}_{n,0})$ denotes a specific value of the initial point x_0 of the phase vector x and $\hat{x} = (\hat{x}_1,...\hat{x}_n)$ is some concrete value of x different from $\hat{x}_0$. Introduce the manifold $X_{t_0} \subseteq R_x^n$ containing all the initial points $x_0 = (x_{1,0},...x_{n,0})$ of the phase vector x. We will call X_{t_0} the manifold of the initial points of the phase vector x.

If we fix the value of the vector of parameter $\xi = \hat{\xi}$ then the differential inclusion (1) turns into the following free dynamic system

$$\frac{dx}{dt} = f\left(t, x; \hat{\xi}\right), \tag{2}$$

where the system (2) can be considered the restriction of the differential inclusion (1) to $T \times R_x^n$.

Let

1. $x = x\left(t; \hat{x}_0, \hat{\xi}\right) = \left(x_1\left(t; \hat{x}_0, \hat{\xi}\right), \ldots, x_n\left(t; \hat{x}_0, \hat{\xi}\right)\right)$ be a particular solution to the system of ordinary differential equations (2) $\forall t \in T$ and at $x_0 = \hat{x}_0 \in X_{t_0}$, which also denotes the specific phase trajectory or orbit just as of the dynamical system (2) so too of the differential inclusion (1) with the concrete initial point of the phase vector $\hat{x}_0 = \left(\hat{x}_{1,0}, \ldots \hat{x}_{n,0}\right) \in X_{t_0}$ and the concrete value $\hat{\xi} = \left(\hat{\xi}_1, \ldots, \hat{\xi}_m\right) \in \Xi$ of the vector of parameters ξ.
2. $x = x\left(t; x_0, \hat{\xi}\right) = \left(x_1\left(t; x_0, \hat{\xi}\right), \ldots, x_n\left(t; x_0, \hat{\xi}\right)\right)$ be a general solution to the dynamical system (2) $\forall x_0 \in X_{t_0}$ and $\hat{\xi} \in \Xi$.
3. $x = x(t; x_0, \xi) = (x_1(t; x_0, \xi), \ldots, x_n(t; x_0, \xi))$ be a general solution to the differential inclusion (1) $\forall x_0 \in X_{t_0}$ and $\forall \xi \in \Xi$.

Since we deal with non-autonomous systems it would be expedient to introduce the integral curve $x_t\left(\hat{x}_0, \hat{\xi}\right) = \left(t, x\left(t; \hat{x}_0, \hat{\xi}\right)\right)$ corresponding to the phase trajectory $x\left(t; \hat{x}_0, \hat{\xi}\right)$. We define the extended phase space or motion space as given below

$$X_t\left(\hat{\xi}\right) = \bigcup_{\forall \hat{x}_0 \in X_{t_0}} x_t\left(\hat{x}_0, \hat{\xi}\right). \tag{3}$$

In the designation $X_t\left(\hat{\xi}\right)$ the symbol $\hat{\xi}$ says that the motion space $X_t\left(\hat{\xi}\right)$ of the dynamic system (2) is just a section of some more total space

$$X_t(\xi) = \bigcup_{\forall \hat{\xi} \in \Xi} X_t\left(\hat{\xi}\right) \tag{4}$$

of the differential inclusion (1) at $\xi = \hat{\xi}$.

Further in our reasoning we will use the terminology and definitions of topological and geometrical objects and constructions according to [2], [3].

In fact, the $x_t\left(\hat{x}_0, \hat{\xi}\right)$ represents a typical fiber of the fiber bundle of integral curves

$$\left\{\Gamma_x\left(t,x;x_0,\xi\right), \pi_x\left(x_0,\xi\right), X_{t_0}\times\Xi, \mathrm{G}_{\hat{x}_0,\hat{\xi}}\left(t,x\right)\right\}, \tag{5}$$

where $\Gamma_x\left(t,x;x_0,\xi\right)=\left\{x_t\left(x_0,\xi\right)=\left(t,x\left(t;x_0,\xi\right)\right),\left(t,x\right)\in T\times R_x^n;\forall\left(x_0,\xi\right)\in X_{t_0}\times\Xi\right\}$ is the total space being a $(n+1;n,m)$-dimensional smooth manifold with the total projection $\pi_x\left(x_0,\xi\right):\Gamma_x\left(t,x;x_0,\xi\right)\to X_{t_0}\times\Xi$, the total base space $X_{t_0}\times\Xi$, the total typical fiber $\Gamma_{\hat{x}_0}\left(t,x;\hat{x}_0,\hat{\xi}\right)=\left\{x_t\left(\hat{x}_0,\hat{\xi}\right)=\left(t,x\left(t;\hat{x}_0,\hat{\xi}\right)\right)\in X_t\left(\hat{\xi}\right)\right\}\cong\pi_x^{-1}\left(\hat{x}_0,\hat{\xi}\right)$ and the one-parameter total trivial structure (Lie) group $\mathrm{G}_{\hat{x}_0,\hat{\xi}}\left(t,x\right)$ acting on the typical fiber $\Gamma_{\hat{x}_0}\left(t,x;\hat{x}_0,\hat{\xi}\right)$.

Definition 1. A fiber bundle is called the fiber metabundle if its typical fiber or/and base space is the total space of another fiber bundle, which is called the fiber subbundle.

Now let us show that the fiber bundle (5) $\left\{\Gamma_x\left(t,x;x_0,\xi\right), \pi_x\left(x_0,\xi\right), X_{t_0}\times\Xi, \mathrm{G}_{\hat{x}_0,\hat{\xi}}\left(t,x\right)\right\}$ can be represented as a fiber metabundle.

Really, it is obvious that the total space (we will call it the total subspace 1) $\Gamma_x\left(t,x;\hat{x}_0,\xi\right)=\left\{x_t\left(\hat{x}_0,\xi\right)=\left(t,x\left(t;\hat{x}_0,\xi\right)\right)\forall\xi\in\Xi\Leftrightarrow\bigcup_{\forall\hat{\xi}\in\Xi}x_t\left(\hat{x}_0,\hat{\xi}\right),\left(t,x\right)\in T\times R_x^n;\hat{x}_0\in X_{t_0}\right\}$ with the projection $\pi_x\left(\hat{x}_0,\xi\right)$ (we will call it the section projection 2), the base space Ξ (we call it the base subspace) and the one-parameter trivial structure (Lie) group $\mathrm{G}_{\hat{x}_0,\hat{\xi}}\left(t,x\right)$ acting on the typical fiber $\Gamma_{\hat{x}_0}\left(t,x;\hat{x}_0,\hat{\xi}\right)=\left\{x_t\left(\hat{x}_0,\hat{\xi}\right)\in X_t\left(\hat{\xi}\right)\right\}\cong\pi_x^{-1}\left(\hat{x}_0,\hat{\xi}\right)$ can be considered a fiber subbundle in the structure of the fiber metabundle (5) as follows

$$\left\{\Gamma_x\left(t,x;\hat{x}_0,\xi\right), \pi_x\left(\hat{x}_0,\xi\right), \Xi, \mathrm{G}_{\hat{x}_0,\hat{\xi}}\left(t,x\right)\right\}. \tag{6}$$

Analogously,

$$\left\{\Gamma_x\left(t,x;x_0,\hat{\xi}\right), \pi_x\left(x_0,\hat{\xi}\right), X_{t_0}, \mathrm{G}_{\hat{x}_0,\hat{\xi}}\left(t,x\right)\right\} \tag{7}$$

can be considered a fiber subbundle in the structure of the fiber metabundle (5) with $\Gamma_x\left(t,x;x_0,\hat{\xi}\right)=\left\{X_t\left(\hat{\xi}\right)=\bigcup_{\forall\hat{x}_0\in X_{t_0}} x_t\left(\hat{x}_0,\hat{\xi}\right),(t,x)\in T\times R_x^n;\hat{\xi}\in\Xi\right\}$ as the total subspace 2, $\pi_x\left(x_0,\hat{\xi}\right)$ as the section projection 1, X_{t_0} as the base subspace and the one-parameter structure group $G_{\hat{x}_0,\hat{\xi}}(t,x)$, which is trivial and acts on the typical fiber $\Gamma_{\hat{x}_0}\left(t,x;\hat{x}_0,\hat{\xi}\right)=$ $=\left\{x_t\left(\hat{x}_0,\hat{\xi}\right)\in X_t\left(\hat{\xi}\right)\right\}\cong\pi_x^{-1}\left(\hat{x}_0,\hat{\xi}\right)$ being equal to the total typical fiber of the (5) and the typical fiber of (6).

Now we are able to describe the fiber bundle (5) $\left\{\Gamma_x(t,x;x_0,\xi),\pi(x_0,\xi),X_{t_0}\times\Xi,G_{x_0,\xi}(t,x)\right\}$ in two ways as

1) a fiber metabundle

$$\left\{\Gamma_x(t,x;x_0,\xi),\pi_x\left(x_0,\hat{\xi}\right),X_{t_0},G_{\hat{x}_0,\xi}(t,x)\right\} \tag{8}$$

with the total metabundle space $\Gamma_x(t,x;x_0,\xi)=\left\{x_t(x_0,\xi)=\left(t,x(t;x_0,\xi)\right),(t,x)\in T\times R_x^n;\right.$ $\left.\forall(x_0,\xi)\in X_{t_0}\times\Xi\right\}$, the section projection 1 $\pi_x\left(x_0,\hat{\xi}\right)$, the total base space X_{t_0}, and the m-parameter structure group $G_{\hat{x}_0,\xi}(t,x)$, which acts on the typical fiber $\Gamma_x(t,x;\hat{x}_0,\xi)=$ $=\left\{X_t^{\xi}(\hat{x}_0)=\bigcup_{\forall\hat{\xi}\in\Xi} x_t\left(\hat{x}_0,\hat{\xi}\right),(t,x)\in T\times R_x^n;\hat{x}_0\in X_{t_0}\right\}\cong\pi_x^{-1}(\hat{x}_0,\xi)$ being the total subspace 1 of the fiber bundle (6), where $X_t^{\xi}(\hat{x}_0)=\bigcup_{\forall\hat{\xi}\in\Xi} x_t\left(\hat{x}_0,\hat{\xi}\right)$ is a bouquet of the integral curves starting from the initial point $\hat{x}_{t_0}=(t_0,\hat{x}_0)$.

2) a fiber metabundle

$$\left\{\Gamma_x(t,x;x_0,\xi),\pi_x(\hat{x}_0,\xi),\Xi,G_{x_0,\hat{\xi}}(t,x)\right\} \tag{9}$$

with the total metabundle space $\Gamma_x(t,x;x_0,\xi)=\{x_t(x_0,\xi)=(t,x(t;x_0,\xi)),(t,x)\in T\times R_x^n;$ $\forall(x_0,\xi)\in X_{t_0}\times\Xi\}$, the section projection 2 $\pi_x(\hat{x}_0,\xi)$, the total base space Ξ and the n-parameter structure group $G_{x_0,\hat{\xi}}(t,x)$, which acts on the typical fiber $\Gamma_x(t,x;x_0,\hat{\xi})=$

$=\left\{X_t(\hat{\xi})=\bigcup_{\forall\hat{x}_0\in X_{t_0}} x_t(\hat{x}_0,\hat{\xi}),(t,x)\in T\times R_x^n;\hat{\xi}\in\Xi\right\}\cong\pi_x^{-1}(x_0,\hat{\xi})$ being the total subspace 2 of the fiber bundle (7).

We need to remind that $X_t(\hat{\xi})=\bigcup_{\forall\hat{x}_0\in X_{t_0}} x_t(\hat{x}_0,\hat{\xi})$ is the extended phase space or motion space corresponding to $\xi=\hat{\xi}\in\Xi$.

Fig. 1 illustrates the structure of this fiber metabundle.

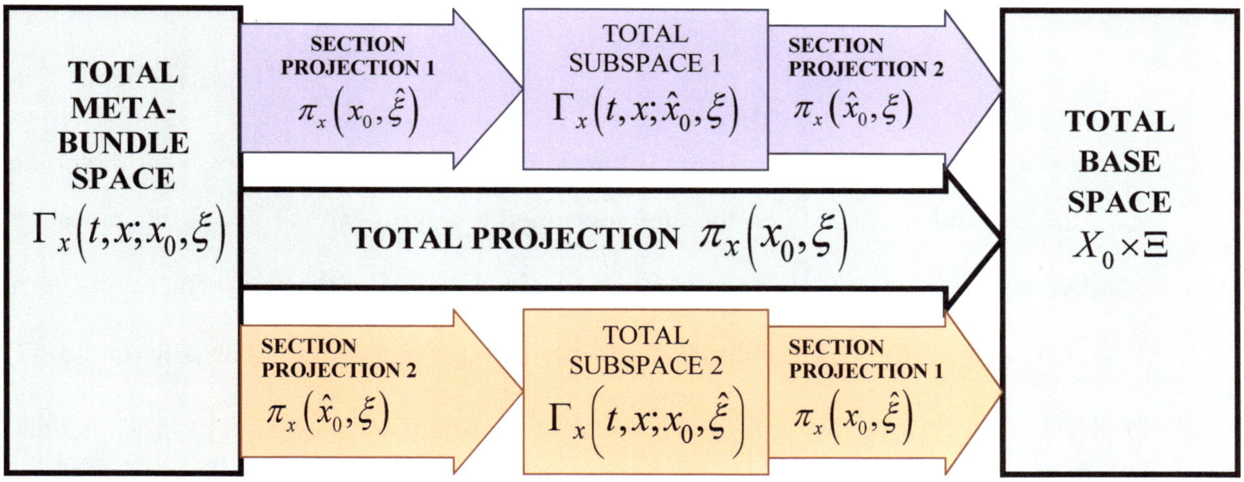

FIG. 1

Thus the fiber metabundle (5) $\{\Gamma_x(t,x;x_0,\xi),X_0\times\Xi,\pi(x_0,\xi),G_{\hat{x}_0,\hat{\xi}}(t,x)\}$ decomposes hierarchically in two ways, namely

i. The inverse section projection 1 of $\pi_x\left(x_0,\hat{\xi}\right)$ binds up all its typical fibers of the form $\Gamma_x\left(t,x;\hat{x}_0,\xi\right)=\left\{X_t^{\xi}\left(\hat{x}_0\right)=\bigcup_{\forall\hat{\xi}\in\Xi}x_t\left(\hat{x}_0,\hat{\xi}\right),\left(t,x\right)\in T\times R_x^n;\hat{x}_0\in X_{t_0}\right\}\cong\pi_x^{-1}\left(\hat{x}_0,\xi\right)$ being the bouquets of the integral curves starting from the same initial points, where x_0 runs over all the points of the base subspace X_{t_0}, in the total metabundle space $\Gamma_x\left(t,x;x_0,\xi\right)$ of the metabouquet or fiber metabundle (5) $\left\{\Gamma_x\left(t,x;x_0,\xi\right),\pi_x\left(x_0,\xi\right),X_{t_0}\times\Xi,\mathrm{G}_{\hat{x}_0,\hat{\xi}}\left(t,x\right)\right\}$. But in its turn, $\Gamma_x\left(t,x;\hat{x}_0,\xi\right)$ is the total subspace of the intermediate-level fiber subbundle (6) $\left\{\Gamma_x\left(t,x;\hat{x}_0,\xi\right),\pi_x\left(\hat{x}_0,\xi\right),\Xi,\mathrm{G}_{\hat{x}_0,\hat{\xi}}\left(t,x\right)\right\}$ with the base subspace Ξ, the section projection 2 $\pi_x\left(\hat{x}_0,\xi\right)$, the structural group $\mathrm{G}_{\hat{x}_0,\hat{\xi}}\left(t,x\right)$ acting on the typical fiber $\Gamma_{\hat{x}_0}\left(t,x;\hat{x}_0,\hat{\xi}\right)=$ $=\left\{x_t\left(\hat{x}_0,\hat{\xi}\right)\in X_t\left(\hat{\xi}\right)\right\}\cong\pi_x^{-1}\left(\hat{x}_0,\hat{\xi}\right)$.

ii. The inverse section projection 2 $\pi_x\left(\hat{x}_0,\xi\right)$ binds up all its typical fibers $\Gamma_x\left(t,x;x_0,\hat{\xi}\right)=\left\{X_t\left(\hat{\xi}\right)=\bigcup_{\forall\hat{x}_0\in X_{t_0}}x_t\left(\hat{x}_0,\hat{\xi}\right),\left(t,x\right)\in T\times R_x^n;\hat{\xi}\in\Xi\right\}\cong\pi_x^{-1}\left(x_0,\hat{\xi}\right)$ being the sheaves of integral curves in the form of motion spaces, where ξ runs over all the points $\hat{\xi}$ of the base subspace Ξ, into the total metabundle space $\Gamma_x\left(t,x;x_0,\xi\right)$ of the metasheaf or fiber metabundle (5) $\left\{\Gamma_x\left(t,x;x_0,\xi\right),\pi_x\left(x_0,\xi\right),X_{t_0}\times\Xi,\mathrm{G}_{\hat{x}_0,\hat{\xi}}\left(t,x\right)\right\}$. But in its turn, $\Gamma_x\left(t,x;x_0,\hat{\xi}\right)$ is the total subspace of the intermediate-level fiber subbundle (7) $\left\{\Gamma_x\left(t,x;x_0,\hat{\xi}\right),\pi_x\left(x_0,\hat{\xi}\right),X_{t_0},\mathrm{G}_{\hat{x}_0,\hat{\xi}}\left(t,x\right)\right\}$ with the section projection 1 $\pi_x\left(x_0,\hat{\xi}\right)$, the base subspace X_{t_0}, the structural group $\mathrm{G}_{\hat{x}_0,\hat{\xi}}\left(t,x\right)$ acting on the typical fiber $\Gamma_{\hat{x}_0}\left(t,x;\hat{x}_0,\hat{\xi}\right)=\left\{x_t\left(\hat{x}_0,\hat{\xi}\right)\in X_t\left(\hat{\xi}\right)\right\}\cong\pi_x^{-1}\left(\hat{x}_0,\hat{\xi}\right)$.

There is one more important thing we need to add to our reasoning. This is the fact that the composition of section projections 1 and 2 is commutative and isomorphic on the fiber metabundle, namely

$$\pi_x\left(x_0,\hat{\xi}\right)\circ\pi_x\left(\hat{x}_0,\xi\right)\cong\pi_x\left(\hat{x}_0,\xi\right)\circ\pi_x\left(x_0,\hat{\xi}\right)\cong\pi_x\left(x_0,\xi\right)\qquad(10)$$

The hierarchical structure of fiber metabundle generated by the inclusion (1) is shown on Fig. 2.

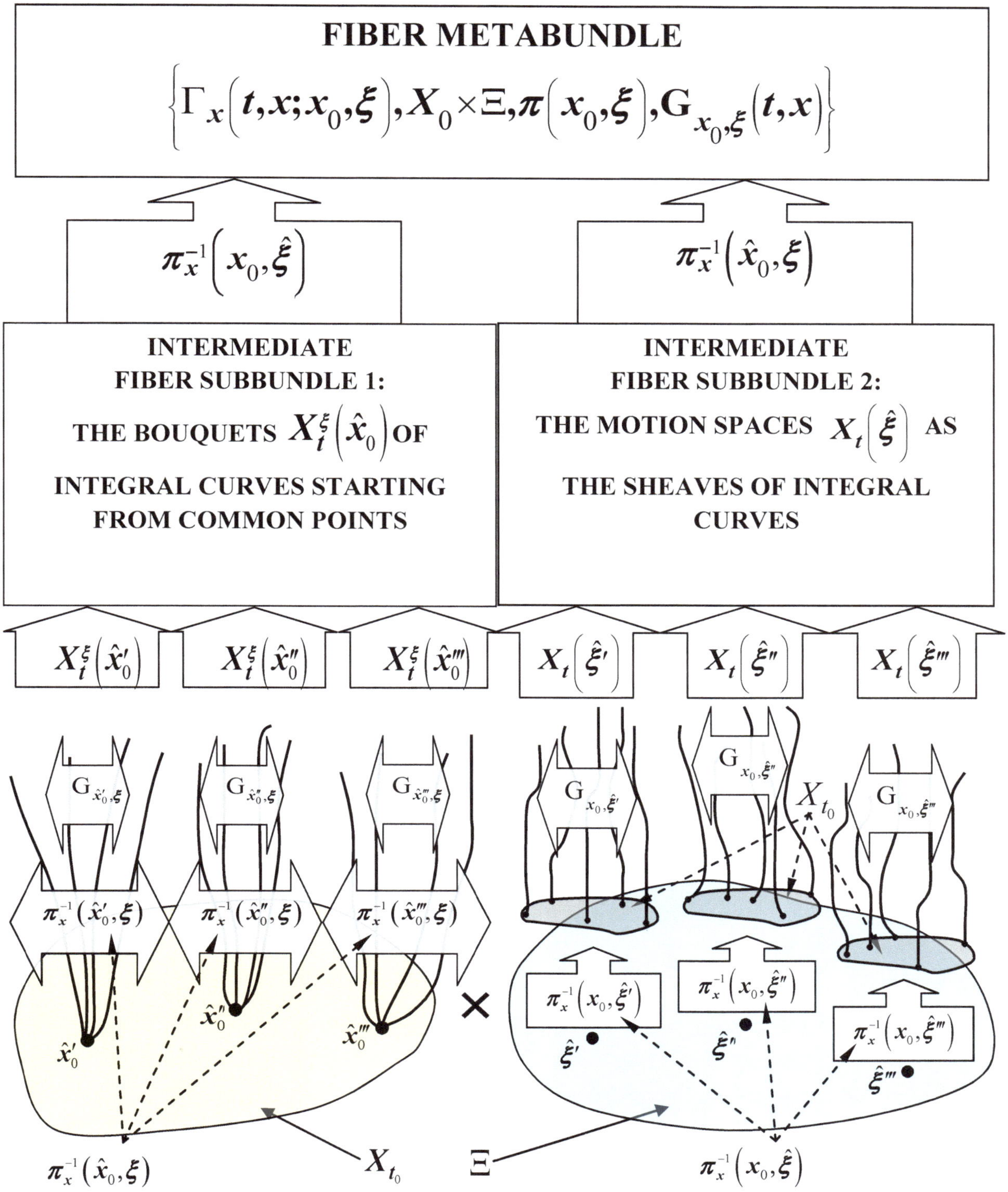
FIBER METABUNDLE
$\{\Gamma_x(t,x;x_0,\xi),X_0\times\Xi,\pi(x_0,\xi),\mathrm{G}_{x_0,\xi}(t,x)\}$
$\pi_x^{-1}(x_0,\hat{\xi})$
$\pi_x^{-1}(\hat{x}_0,\xi)$
INTERMEDIATE FIBER SUBBUNDLE 1:
THE BOUQUETS $X_t^\xi(\hat{x}_0)$ OF
INTEGRAL CURVES STARTING FROM COMMON POINTS
INTERMEDIATE FIBER SUBBUNDLE 2:
THE MOTION SPACES $X_t(\hat{\xi})$ AS
THE SHEAVES OF INTEGRAL CURVES
$X_t^\xi(\hat{x}_0')$
$X_t^\xi(\hat{x}_0'')$
$X_t^\xi(\hat{x}_0''')$
$X_t(\hat{\xi}')$
$X_t(\hat{\xi}'')$
$X_t(\hat{\xi}''')$
$\mathrm{G}_{\hat{x}_0',\xi}$
$\mathrm{G}_{\hat{x}_0'',\xi}$
$\mathrm{G}_{\hat{x}_0''',\xi}$
$\mathrm{G}_{x_0,\hat{\xi}'}$
$\mathrm{G}_{x_0,\hat{\xi}''}$
$\mathrm{G}_{x_0,\hat{\xi}'''}$
X_{t_0}
$\pi_x^{-1}(\hat{x}_0',\xi)$
$\pi_x^{-1}(\hat{x}_0'',\xi)$
$\pi_x^{-1}(\hat{x}_0''',\xi)$
$\pi_x^{-1}(x_0,\hat{\xi}')$
$\pi_x^{-1}(x_0,\hat{\xi}'')$
$\pi_x^{-1}(x_0,\hat{\xi}''')$
$\hat{x}_0'$
$\hat{x}_0''$
$\hat{x}_0'''$
$\hat{\xi}'$
$\hat{\xi}''$
$\hat{\xi}'''$
×
$\pi_x^{-1}(\hat{x}_0,\xi)$
X_{t_0}
Ξ
$\pi_x^{-1}(x_0,\hat{\xi})$

FIG. 2

Suppose the inclusion has a complete set of independent parametric first integrals

$$g(t,x;\xi)=\left(g_1(t,x;\xi),...,g_n(t,x;\xi)\right)\in C^{r+1}\ , \tag{11}$$

which are dependent of the vector of parameters ξ.

Set up the following vector equation

$$g(t,x;\xi)=c\Leftrightarrow\left\{g_i(t,x;\xi)=c_i,\forall c_i=g_i(t_0,x_0;\xi)\in R^1,i=(1,...,n)\right\}, \tag{12}$$

where $c=(c_1,...,c_n)=g(t_0,x_0;\xi)=\left(g_1(t_0,x_0;\xi),...,g_n(t_0,x_0;\xi)\right)$, $(x_0,\xi)\in X_{t_0}\times\Xi$. At some fixed $c=\hat{c}$ and $\xi=\hat{\xi}$ the equation (12) defines the set of n n-dimensional invariant manifolds $\left\{M_{x_1}(\hat{c}_1,\hat{\xi}),....,M_{x_n}(\hat{c}_n,\hat{\xi})\right\}$, where

$$M_{x_i}(\hat{c}_i,\hat{\xi})=\left\{g_i(t,x;\hat{\xi})=\hat{c}_i,(t,x)\in T\times R_x^n;\hat{\xi}\in\Xi\right\}. \tag{13}$$

In its turn, the following vector equation for x_0, which we also consider parametric variables,

$$\hat{c}=g(t_0,x_0;\hat{\xi}) \tag{14}$$

defines the set of n $(n-1)$-dimensional submanifolds $\left\{L_{x_i}(\hat{c}_i,\hat{\xi})\right\}_{i=1}^n\subset X_{t_0}$ of the corresponding above-mentioned manifolds $\left\{M_{x_i}(\hat{c}_i,\hat{\xi})\right\}_{i=1}^n$ that belong to $T\times R_x^n$. The former ones we obtain by means of the intersection the latter ones with the hyperplane $\{t=t_0\}$ that is

$$\left\{L_{x_i}(\hat{c}_i,\hat{\xi})\right\}_{i=1}^n=\left\{M_{x_i}(\hat{c}_i,\hat{\xi})\cap\{t=t_0\}\right\}_{i=1}^n \tag{15}$$

where $\left\{L_{x_i}(\hat{c}_i,\hat{\xi})=\left\{\hat{c}_i=g_i(t_0,x_0;\hat{\xi}),x_0\in X_{t_0};\hat{\xi}\in\Xi\right\}\right\}_{i=1}^n$.

First, we assume that each i-component equation of the vector equation

$$\left\{g_i(t,x;\xi)=c_i\big|_{c_i=g_i(t_0,x_0;\xi)}\right\}_{i=1}^n, \tag{16}$$

obtained from (12), can be solved for the corresponding component x_i in a unique manner

$$\left\{ g_i\left(t,x;\xi\right)=c_i\big|_{c_i=g_i\left(t_0,x_0;\xi\right)} \right\}_{i=1}^{n} \Rightarrow \left| \begin{matrix} g_i\left(t,x;\xi\right)=c_i \Rightarrow \\ \Rightarrow x_i=\varphi_i\left(t,x^i;c_i,\xi\right) \end{matrix} \right| \Rightarrow \left\{ x_i=\varphi_i\left(t,x^i;c_i,\xi\right)\big|_{c_i=g_i\left(t_0,x_0;\xi\right)} \right\}_{i=1}^{n} = \tag{17}$$

$$=\left\{ x_i=\varphi_i\left(t,x^i;g_i\left(t_0,x_0;\xi\right),\xi\right)\right\}_{i=1}^{n}=\left\{ x_i=\tilde{\varphi}_i\left(t,x^i;t_0,x_0,\xi\right)\right\}_{i=1}^{n},$$

where $x^i=\left(x_1,...,x_{i-1},x_{i+1},...,x_n\right)$, $\tilde{\varphi}_i\left(t,x^i;t_0,x_0,\xi\right)\in C^{r+1}$, $\left(t,x^i\right)\in R^n_{t,x^i}$, $\left(x_0,\xi\right)\in X_{t_0}\times\Xi$.

In vector form this condition looks as follows

$$x=\tilde{\varphi}\left(t,x;t_0,x_0,\xi\right), \tag{18}$$

where $\tilde{\varphi}\left(t,x;t_0,x_0,\xi\right)=\left(\tilde{\varphi}_1\left(t,x^1;t_0,x_0,\xi\right),...,\tilde{\varphi}_k\left(t,x^k;t_0,x_0,\xi\right),...,\tilde{\varphi}_n\left(t,x^n;t_0,x_0,\xi\right)\right)$,

$k\in\left(2,...,n-1\right)$, $x^1=\left(x_2,...,x_n\right)$, $x^n=\left(x_1,...,x_{n-1}\right)$.

Remark 1. *The expression from (17) means that the* 1*-codimensional or* n*-dimensional invariant manifolds* $\left\{M_{x_i}\left(\hat{c}_i,\hat{\xi}\right)\right\}_{i=1}^{n}$ *can be equally represented by explicit and implicit functional equations, namely*

$$\left\{ M_{x_i}\left(\hat{c}_i,\hat{\xi}\right)=\left\{ g_i\left(t,x;\hat{\xi}\right)=\hat{c}_i\big|_{\hat{c}_i=g_i\left(t_0,\hat{x}_0;\hat{\xi}\right)} \Leftrightarrow x_i=\left\{ \begin{matrix} =\varphi_i\left(t,x^i;\hat{c}_i,\hat{\xi}\right) \\ =\tilde{\varphi}_i\left(t,x^i;t_0,\hat{x}_0,\hat{\xi}\right) \end{matrix} \right\}, \left(t,x\right)\in T\times R^n_x;\hat{\xi}\in\Xi \right\} \right\}_{i=1}^{n}. \tag{19}$$

As to the explicit expressions representing the 2-codimensional or $\left(n-1\right)$*-dimensional submanifolds* $\left\{L_{x_i}\left(\hat{c}_i,\hat{\xi}\right)\right\}_{i=1}^{n}$ *of the former ones we have to put* $\left(t=t_0,x=x_0\right)$ *in the left-hand side of the equations (16) and solve each of them for* $x_{i,0}$ *in the corresponding order, where* $i=\left(1,...,n\right)$*, then fix the value of* c *and* ξ*, that is set* $c_i=\hat{c}_i$, $\xi=\hat{\xi}$ *as it is shown below*

$$\left\{ g_i\left(t_0,x_0;\xi\right)=c_i \Rightarrow x_{i,0}=\varphi_i\left(t_0,x_0^i,\xi;c_i\right) \Rightarrow \left| \begin{matrix} c_i=\hat{c}_i \\ \xi=\hat{\xi} \end{matrix} \right| \Rightarrow x_{i,0}=\varphi_i\left(t_0,x_0^i,\hat{\xi};\hat{c}_i\right) \right\}_{i=1}^{n}. \tag{20}$$

Finally we obtain

$$\left\{ L_{x_i}\left(\hat{c}_i,\hat{\xi}\right)=\left\{ \hat{c}_i=g_i\left(t_0,x_0;\hat{\xi}\right) \Leftrightarrow x_{i,0}=\varphi_i\left(t_0,x_0^i,\hat{\xi};\hat{c}_i\right), x_0\in X_{t_0};\hat{\xi}\in\Xi \right\} \right\}_{i=1}^{n}. \tag{21}$$

Second, assume that the vector equation consisting of n of the component equation (20) can be solved for x_0 in a unique manner too, namely

$$\begin{pmatrix} \left\{g_i\left(t_0,x_0;\xi\right)=c_i\right\}_{i=1}^n \\ \Updownarrow \\ g\left(t_0,x_0;\xi\right)=c \end{pmatrix} \Rightarrow x_0=\psi\left(t_0,c,\xi\right), \tag{22}$$

where the smooth vector function $\psi\left(t_0,c,\xi\right)=\left(\psi_1\left(t_0,c,\xi\right),\ldots,\psi_n\left(t_0,c,\xi\right)\right)$ defines the diffeomorphism $\psi:R^n\times\Xi\to X_{t_0}$ for $t_0\in \mathrm{Fr}T$.

Third, let us introduce a new fiber bundle with the projection based on the relation (22)

$$\psi:R^n\times\Xi\to\left\{\bigcup_{\forall\hat{\xi}\in\Xi}\left(\bigcup_{\forall\hat{c}_i\in R^1}L_{x_i}\left(\hat{c}_i,\hat{\xi}\right)\right)\right\}_{i=1}^n=X_{t_0}. \tag{23}$$

Its total space is $X_{t_0}=\left\{\bigcup_{\forall\xi\in\Xi}\left(\bigcup_{\forall c_i\in R^1}L_{x_i}\left(c_i,\xi\right)\right)\right\}_{i=1}^n$, the base space is $R^n\times\Xi$, the typical fiber is

$$\left\{L_{x_i}\left(\hat{c}_i,\hat{\xi}\right)=\left\{\hat{c}_i=g_i\left(t_0,x_0;\hat{\xi}\right)\Rightarrow x_{i,0}=\varphi_i\left(t_0,x_0^{i},\hat{\xi};\hat{c}_i\right),x_0\in X_{t_0};\hat{\xi}\in\Xi\right\}\right\}_{i=1}^n. \tag{24}$$

This means that we can construct *the collection of quotient spaces* $\left\{X_{t_0}/L_{x_i}\left(\hat{c}_i,\hat{\xi}\right)\right\}_{i=1}^n$ for X_{t_0} through the homeomorphism ψ as an equivalence relation acting on X_{t_0}. Further in our reasoning if the vector of parameter ξ is fixed and equals some $\hat{\xi}$ it is necessary to replace the total base space X_{t_0} with the total base space $X_{t_0}/L_{x_i}\left(\hat{c}_i,\hat{\xi}\right)$ for each specific component i in the fiber bundles generated by the equation (12).

Consider a fiber bundle determined by the equation (12), namely

$$\left\{\Gamma_{x_i}\left(t,x;c_i,\xi\right),\pi_{x_i}\left(c_i,\xi\right),R^1\times\Xi,\mathrm{G}_{\hat{c}_i,\hat{\xi}}^{x_i}\left(t,x\right)\right\},\quad n\in\left(1,\ldots,n\right) \tag{25}$$

with the total space $\Gamma_{x_i}\left(t,x;c_i,\xi\right)=\left\{g_i\left(t,x;\xi\right)=c_i,\left(t,x\right)\in T\times R_x^n;\forall\left(c_i,\xi\right)\in R^1\times\Xi\right\}$, the total projection $\pi_{x_i}\left(c_i,\xi\right):\Gamma_{x_i}\left(t,x;c_i,\xi\right)\to R^1\times\Xi$, the total base space $R^1\times\Xi$, the total typical fiber

$\Gamma_{x_i}\left(t,x;\hat{c}_i,\hat{\xi}\right)=M_{x_i}\left(\hat{c}_i,\hat{\xi}\right)=\left\{g_i\left(t,x;\hat{\xi}\right)=\hat{c}_i,(t,x)\in T\times R_x^n;\left(\hat{c}_i,\hat{\xi}\right)\in R^1\times\Xi\right\}=\pi_{x_i}^{-1}\left(\hat{c}_i,\hat{\xi}\right)$ and the $(n-1)$-parameter structure Lie group $G_{\hat{c}_i,\hat{\xi}}^{x_i}(t,x)$ acting on it, where the elements of this group are the integral curves $x_t\left(x_0,\hat{\xi}\right)\in\Gamma_{x_i}\left(t,x;\hat{c}_i,\hat{\xi}\right)=M_{x_i}\left(\hat{c}_i,\hat{\xi}\right)$.

Obviously, the fiber bundle (25) can be considered a fiber metabundle with the typical fiber represented by the total spaces of the two other fiber subbundles, namely

1) the fiber subbundle in the structure of the fiber metabundle (25)

$$\left\{\Gamma_{x_i}\left(t,x;\hat{x}_0,\xi\right),\pi_{x_i}\left(\hat{x}_0,\xi\right)\xleftrightarrow{\hat{c}_i=g_i(t_0,\hat{x}_0;\xi)}\pi_{x_i}\left(\hat{c}_i,\xi\right),\Xi,G_{\hat{x}_0,\hat{\xi}}^{x_i}(t,x)\right\} \tag{26}$$

with the total subspace

$\Gamma_{x_i}\left(t,x;\hat{x}_0,\xi\right)=F_{x_i}^{\xi}\left(\hat{x}_0\right)=\left\{\bigcup_{\forall\hat{\xi}\in\Xi}\left\{x_i=\tilde{\varphi}_i\left(t,x^i;t_0,\hat{x}_0,\hat{\xi}\right)\right\},(t,x)\in T\times R_x^n;\hat{x}_0\in X_{t_0}\right\}$, the section projection 2 $\left\{\pi_{x_i}\left(\hat{x}_0,\xi\right)\xleftrightarrow{\hat{c}_i=g_i(t_0,\hat{x}_0;\xi)}\pi_{x_i}\left(\hat{c}_i,\xi\right)\right\}:\Gamma_{x_i}\left(t,x;\hat{x}_0,\xi\right)\to\Xi$, the base subspace Ξ, the typical fiber

$\Gamma_{x_i}\left(t,x;\hat{x}_0,\hat{\xi}\right)\xleftrightarrow{\hat{c}_i=g_i\left(t_0,\hat{x}_0;\hat{\xi}\right)}M_{x_i}\left(\hat{c}_i,\hat{\xi}\right)=\left\{x_i=\tilde{\varphi}_i\left(t,x^i;t_0,\hat{x}_0,\hat{\xi}\right),(t,x)\in T\times R_x^n;\hat{x}_0\in X_{t_0},\hat{\xi}\in\Xi\right\}=$

$=\pi_{x_i}^{-1}\left(\hat{x}_0,\hat{\xi}\right)\xleftrightarrow{\hat{c}_i=g_i\left(t_0,\hat{x}_0;\hat{\xi}\right)}\pi_{x_i}^{-1}\left(\hat{c}_i,\hat{\xi}\right)$ and the $(n-1)$-parameter structure group $G_{\hat{x}_0,\hat{\xi}}^{x_i}(t,x)$ acting on it, where the elements of this group are integral curves $x_t\left(x_0,\hat{\xi}\right)\in\Gamma_{x_i}\left(t,x;\hat{c}_i,\hat{\xi}\right)=M_{x_i}\left(\hat{c}_i,\hat{\xi}\right)$.

2) the fiber subbundle in the structure of the fiber metabundle (25)

$$\left\{\Gamma_{x_i}\left(t,x;x_0,\hat{\xi}\right),\pi_{x_i}\left(c_i,\hat{\xi}\right),R^1,G_{\hat{x}_0,\hat{\xi}}^{x_i}(t,x)\right\} \tag{27}$$

with the total subspace $\Gamma_{x_i}\left(t,x;x_0,\hat{\xi}\right)=F_{x_i}\left(\hat{\xi}\right)=$

$=\left\{\bigcup_{\forall\hat{c}_i\in R^1}\left\{g_i\left(t,x;\hat{\xi}\right)=\hat{c}_i\Big|_{\hat{c}_i=g_i\left(t_0,x_0;\hat{\xi}\right)}\Leftrightarrow x_i=\tilde{\varphi}_i\left(t,x^i;t_0,x_0,\hat{\xi}\right)\right\},(t,x)\in T\times R_x^n;\hat{\xi}\in\Xi\right\}$, the section projection 1 $\pi_{x_i}\left(c_i,\hat{\xi}\right):\Gamma_{x_i}\left(t,x;x_0,\hat{\xi}\right)\to R^1$, the base subspace R^1, the typical fiber

$\Gamma_{x_i}\left(t,x;\hat{x}_0,\hat{\xi}\right)\xleftrightarrow{\hat{c}_i=g_i\left(t_0,\hat{x}_0;\hat{\xi}\right)}M_{x_i}\left(\hat{c}_i,\hat{\xi}\right)=\left\{g_i\left(t,x;\hat{\xi}\right)=\hat{c}_i\Big|_{\hat{c}_i=g_i\left(t_0,\hat{x}_0;\hat{\xi}\right)}\Leftrightarrow x_i=\tilde{\varphi}_i\left(t,x^i;t_0,\hat{x}_0,\hat{\xi}\right),\right.$

$\left.(t,x)\in T\times R_x^n;\hat{c}_i\in R^1,\hat{\xi}\in\Xi\right\}=\pi_{x_i}^{-1}\left(\hat{c}_i,\hat{\xi}\right)$ and the $(n-1)$-parameter structure group $G_{\hat{x}_0,\hat{\xi}}^{x_i}(t,x)$

acting on it, where the elements of this group are integral curves $x_t\left(x_0,\hat{\xi}\right)\in\Gamma_{x_i}\left(t,x;\hat{c}_i,\hat{\xi}\right)$. In fact this fiber bundle is a 1-codimensional foliation.

Remark 2. *Now we will show why the dimension of the structure Lie group* $\mathrm{G}^{x_i}_{\hat{x}_0,\hat{\xi}}\left(t,x\right)$ *is* $\left(n-1\right)$. *Consider the vector equation*

$$g^i\left(t,x;\hat{\xi}\right)=c^i$$

for $x^i=\left(x_1,\ldots,x_{i-1},x_{i+1},\ldots,x_n\right)$, *where* $c^i=\left(c_1,\ldots,c_{i-1},c_{i+1},\ldots,c_n\right)$, $g^i\left(t,x;\hat{\xi}\right)=\left(g_1\left(t,x;\hat{\xi}\right),\ldots,g_{i-1}\left(t,x;\hat{\xi}\right),g_{i+1}\left(t,x;\hat{\xi}\right),\ldots,g_n\left(t,x;\hat{\xi}\right)\right)$. *Now plug* $x_i=\varphi_i\left(t,x^i;\hat{c}_i,\hat{\xi}\right)$ *from (19) in it*

$$g^i\left(t,x^i,\varphi_i\left(t,x^i;\hat{c}_i,\hat{\xi}\right);\hat{\xi}\right)=\tilde{g}^i\left(t,x^i;\hat{c}_i,\hat{\xi}\right)=c^i.$$

Solve the obtained vector equation for x^i. *We receive*

$$x^i=p^i\left(t;\hat{c}_i,\hat{\xi},c^i\right),$$

where $p^i\left(t;\hat{c}_i,\hat{\xi},c^i\right)=\left(p_1\left(t;\hat{c}_i,\hat{\xi},c^i\right),\ldots,p_{i-1}\left(t;\hat{c}_i,\hat{\xi},c^i\right),p_{i+1}\left(t;\hat{c}_i,\hat{\xi},c^i\right),\ldots,p_n\left(t;\hat{c}_i,\hat{\xi},c^i\right)\right)$.

We should not forget that $\hat{c}_i,\hat{\xi}$ *are some concrete numbers and in general they can be left out in formulas. The last vector equation describes the orthogonal projection of the family of integral curves given as follows*

$$x=\left(p_1\left(t;\hat{c}_i,\hat{\xi},c^i\right),\ldots,p_{i-1}\left(t;\hat{c}_i,\hat{\xi},c^i\right),\varphi_i\left(t,p^i\left(t;\hat{c}_i,\hat{\xi},c^i\right);\hat{c}_i,\hat{\xi}\right),p_{i+1}\left(t;\hat{c}_i,\hat{\xi},c^i\right),\ldots,p_n\left(t;\hat{c}_i,\hat{\xi},c^i\right)\right)$$

in $R^{n-1}_{x^i}$, *which depends on* $\left(n-1\right)$ *parameters* c^i *and compose the* n*-dimensional manifold* $\Gamma_{x_i}\left(t,x;\hat{c}_i,\hat{\xi}\right)=M_{x_i}\left(\hat{c}_i,\hat{\xi}\right)$.

Remark 3. *Once again it is very important to emphasize that despite the fact that* x_0 *and* ξ *are considered the vectors of parameters, they do not have the equal rights in the hierarchical structure of the fiber metabundle (25) generated by the complete set of the first integrals (11) (or (12)) of the differential inclusion (1). The vector of parameters* $\xi\in\Xi$ *has the first-grade or absolute independence. Meanwhile* $x_0\in X_{t_0}$ *has the second-grade independence since* $c=\left(c_1,\ldots,c_n\right)$ *breaks* X_{t_0} *in the equivalence classes* $L_{x_i}\left(\hat{c}_i,\hat{\xi}\right)\cong\psi_i\left(\hat{c}_i,\hat{\xi}\right)$ *creating the quotient* *space* $X_{t_0} / L_{x_i}\left(\hat{c}_i, \hat{\xi}\right)$ *from* X_{t_0}. *It is clearly seen from (22). Fixing* x_0 *we choose some concrete point* $\hat{x}_0$ *of phase space and are able to investigate the bouquet of leaves having the point* $\hat{x}_0$ *as common one of different 1-codimensional foliations corresponding to the different points of* $\hat{\xi} \in \Xi$. *Fixing* ξ *we pick out some specific 1-codimensional foliation being the sheaf of 1-codimensional manifolds having the latitude to investigate the properties of its leaves. See Fig.3.*

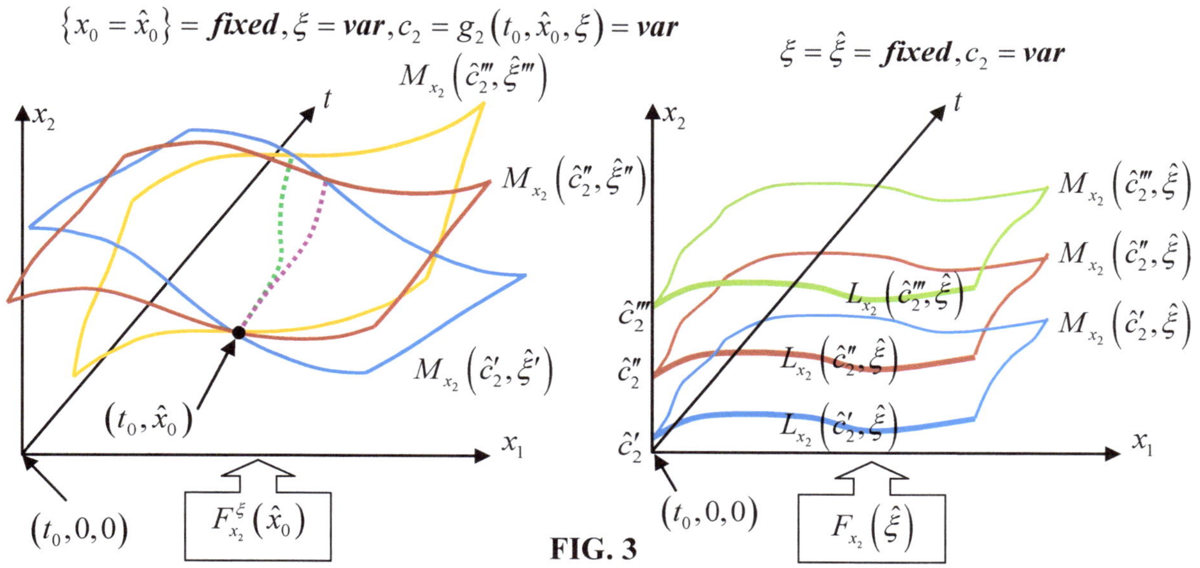

FIG. 3

The situation with x_0 and ξ reminds the one with commuting and anticommuting variables in supermanifolds [4] and can be investigated additionally if the components of ξ are anticommuting ones.

On the analogy of the fiber bundle (5) we can present the fiber bundle (25) in two ways as follows

1. a fiber metabundle

$$\left\{\left\{\begin{matrix} \Gamma_{x_i}(t, x; x_0, \xi) \xleftrightarrow{c_i = g_i(t_0, x_0; \hat{\xi})} \Gamma_{x_i}(t, x; c_i, \xi) \\ \pi_{x_i}\left(x_0, \hat{\xi}\right) \xleftrightarrow{c_i = g_i(t_0, x_0; \hat{\xi})} \pi_{x_i}\left(c_i, \hat{\xi}\right) \end{matrix}\right\}, X_{t_0}, \mathrm{G}_{\hat{x}_0, \xi}^{x_i}(t, x)\right\} \tag{28}$$

with the total metabundle space $\Gamma_{x_i}(t, x; x_0, \xi) \xleftrightarrow{c_i = g_i(t_0, x_0; \hat{\xi})} \Gamma_{x_i}(t, x; c_i, \xi) =$

$$= \left\{ g_i(t, x; \xi) = c_i \big|_{c_i = g_i(t_0, x_0; \xi)} \Leftrightarrow x_i = \tilde{\varphi}_i\left(t, x^i; t_0, x_0, \xi\right), (t, x) \in T \times R_x^n; \forall (x_0, \xi) \in X_{t_0} \times \Xi \right\},$$

the section projection 1 $\pi_{x_i}\left(x_0, \hat{\xi}\right) \xleftrightarrow{c_i = g_i(t_0, x_0; \hat{\xi})} \pi_{x_i}\left(c_i, \hat{\xi}\right)$, the total base space X_{t_0} and the m-parameter structure group $G^{x_i}_{\hat{x}_0,\xi}(t,x)$ acting on the typical fiber

$$\Gamma_{x_i}(t,x;\hat{x}_0,\xi)=F^{\xi}_{x_i}(\hat{x}_0)=\left\{\bigcup_{\forall\xi\in\Xi}\left\{x_i=\tilde{\varphi}_i\left(t,x^i;t_0,\hat{x}_0,\xi\right)\right\},(t,x)\in T\times R^n_x;\hat{x}_0\in X_{t_0}\right\}=\pi^{-1}_{x_i}(\hat{x}_0,\xi)$$

being the total subspace of the fiber subbundle (26), where $F^{\xi}_{x_i}(\hat{x}_0)=\bigcup_{\forall\xi\in\Xi}\left\{x_i=\tilde{\varphi}_i\left(t,x^i;t_0,\hat{x}_0,\xi\right)\right\}$ is a bouquet of the 1-codimensional invariant manifolds having the initial point $\hat{x}_{t_0}=(t_0,\hat{x}_0)$ of the integral curves x_t corresponding to each $\hat{\xi}\in\Xi$ as common one.

2. a fiber metabundle

$$\left\{\Gamma_{x_i}(t,x;x_0,\xi)\xleftrightarrow{c_i=g_i\left(t_0,x_0;\hat{\xi}\right)}\Gamma_{x_i}(t,x;c_i,\xi),\pi_{x_i}(\hat{c}_i,\xi),\Xi,G^{x_i}_{x_0,\hat{\xi}}(t,x)\right\} \tag{29}$$

with the total metabundle space $\Gamma_{x_i}(t,x;x_0,\xi)\xleftrightarrow{c_i=g_i\left(t_0,x_0;\hat{\xi}\right)}\Gamma_{x_i}(t,x;c_i,\xi)=$ $=\left\{g_i(t,x;\xi)=c_i\big|_{\hat{c}_i=g_i(t_0,x_0;\xi)}\Leftrightarrow x_i=\tilde{\varphi}_i\left(t,x^i;t_0,x_0,\xi\right),(t,x)\in T\times R^n_x;\forall(x_0,\xi)\in X_{t_0}\times\Xi\right\}$, the section projection 2 $\pi_{x_i}(\hat{c}_i,\xi)$, the total base space Ξ, and the n-parameter structure group $G^{x_i}_{x_0,\hat{\xi}}(t,x)$ that acts on the typical fiber $\Gamma_{x_i}\left(t,x;c_i,\hat{\xi}\right)=\Gamma_{x_i}\left(t,x;x_0,\hat{\xi}\right)=F_{x_i}\left(\hat{\xi}\right)=$

$$=\left\{\bigcup_{\forall c_i\in R^1}\left\{g_i\left(t,x;\hat{\xi}\right)=c_i\big|_{c_i=g_i\left(t_0,x_0;\hat{\xi}\right)}\Leftrightarrow x_i=\tilde{\varphi}_i\left(t,x^i;t_0,x_0,\hat{\xi}\right)\right\},(t,x)\in T\times R^n_x;\hat{\xi}\in\Xi\right\}=\pi^{-1}_{x_i}\left(c_i,\hat{\xi}\right),$$

which is in its turn the total subspace 2 of the fiber bundle (27). We suppose here that $F_{x_i}\left(\hat{\xi}\right)=\bigcup_{\forall c_i\in R^1}\left\{g_i\left(t,x;\hat{\xi}\right)=c_i\big|_{c_i=g_i\left(t_0,x_0;\hat{\xi}\right)}\right\}$ is a foliation of the invariant manifolds of 1-codimension being the union of all the sections of the first integral $g_i(t,x;\xi)$ created by hyperplanes $c_i=\hat{c}_i\ \forall\hat{c}_i\in R^1$ at some fixed point $\xi=\hat{\xi}\in\Xi$.

3. PROCEDURE OF FLATTENING TOTAL SPACES OF FIBER BUNDLES

3.1. The initial system and the cascade of flattening diffeomorphisms

Now once again we indicate the whole initial system that will be subject to the procedure of sequential transformations by means of the cascade of flattening diffeomorphisms. The one is

$$\left\{ \left\{ \Gamma_{x_i}\left(t,x;c_i,\xi\right), \pi_{x_i}\left(c_i,\xi\right), R^1 \times \Xi, G^{x_i}_{\hat{c}_i,\hat{\xi}}\left(t,x\right) \right\}_{i=1}^{n}; \frac{dx}{dt} \in \bigcup_{\xi \in \Xi} f\left(t,x;\xi\right) \right\}, \tag{30}$$

where $f\left(t,x;\xi\right) = \left(f_1\left(t,x;\xi\right),...,f_n\left(t,x;\xi\right)\right) \in C^r, g\left(t,x;\xi\right) = C^{r+1}$. As it can be seen from (30) the whole initial system is composed of the differential inclusion (1) and the set of all the fiber metabundles (25) for all the components of the phase vector x. The cascade of sequential flattening diffeomorphisms is defined as

$$\theta = \left\{ \begin{array}{l} \underbrace{\begin{array}{l}\theta_1 = \left\{ x_1 = y_1 + \varphi_1\left(t,x_2,...,x_n;c_1,\xi\right) \right\} : \Gamma_{x_1}\left(t,x;c_1,\xi\right) \to \\ \to \Gamma^1_{x_1}\left(t,y_1,x_2,...,x_n;c_1,\xi\right)\left\{y_1 = 0\right\} \subset R^{2n+m+1}\end{array}} \\ \Downarrow \\ \underbrace{\begin{array}{l}\theta_2 = \left\{ x_2 = y_2 + \varphi^1_2\left(t,y_1,x_3,...,x_n;c_1,c_2,\xi\right) \right\} : \Gamma^1_{x_2}\left(t,y_1,x_2,...,x_n;c_1,\xi\right) \to \\ \to \Gamma^2_{x_2}\left(t,y_1,y_2,x_3,...,x_n;c_1,c_2,\xi\right) = \left\{y_2 = 0\right\} \subset R^{2n+m+1}\end{array}} \\ \Downarrow \\ \underbrace{\begin{array}{l}\theta_3 = \left\{ x_3 = y_3 + \varphi^2_3\left(t,y_1,y_2,x_4,...,x_n;c_1,c_2,c_3,\xi\right) \right\} : \\ \Gamma^2_{x_3}\left(t,y_1,y_2,x_3,...,x_n;c_1,c_2,\xi\right) \to \Gamma^3_{x_3}\left(t,y_1,y_2,y_3,x_4,...,x_n;c_1,c_2,c_3,\xi\right) = \\ = \left\{y_3 = 0\right\} \subset R^{2n+m+1}\end{array}} \\ \Downarrow \\ \cdots\cdots\cdots\cdots\cdots\cdots\cdots\cdots \\ \underbrace{\begin{array}{l}\theta_k = \left\{ x_k = y_k + \varphi^{k-1}_k\left(t,y_1,...,y_{k-1},x_{k+1},...,x_n;c_1,...,c_k,\xi\right) \right\} : \\ \Gamma^k_{x_{k-1}}\left(t,y_1,...,y_{k-1},x_k,...,x_n;c_1,...,c_{k-1},\xi\right) \to \\ \to \Gamma^k_{x_k}\left(t,y_1,...,y_k,x_{k+1},...,x_n;c_1,...,c_k,\xi\right) = \left\{y_k = 0\right\} \subset R^{2n+m+1}\end{array}} \\ \Downarrow \\ \cdots\cdots\cdots\cdots\cdots\cdots\cdots\cdots \\ \theta_n = \left\{ x_n = y_n + \varphi^{n-1}_n\left(t,y_1,...,y_{n-1};c,\xi\right) \right\} : \Gamma^n_{x_{n-1}}\left(t,y_1,...,y_{n-1},x_n;c_1,...,c_{n-1},\xi\right) \to \\ \to \Gamma^n_{x_n}\left(t,y;c,\xi\right) = \left\{y_n = 0\right\} \subset R^{2n+m+1} \end{array} \right\}, \tag{31}$$

where $\theta = \left(\theta_1 \Rightarrow \theta_2 \Rightarrow \theta_3 \Rightarrow ... \Rightarrow \theta_{k-1} \Rightarrow \theta_k \Rightarrow ... \Rightarrow \theta_{n-1} \Rightarrow \theta_n\right)$. In fact, the flattening diffeomorphisms are the orthogonal projections of geometrical-topological objects embedded namely, $\left\{\Gamma_{x_i}\left(t,x;c_i,\xi\right)\right\}_{i=1}^{n}$ in $R_{t,x,c,\xi}^{2n+m+1}$ onto its $2n+m$-dimensional subspaces $\left\{R_{t,x^1,c,\xi}^{2n+m}, R_{t,x^2,c,\xi}^{2n+m}, R_{t,x^3,c,\xi}^{2n+m},\ldots, R_{t,x^{k-1},c,\xi}^{2n+m}, R_{t,x^k,c,\xi}^{2n+m},\ldots, R_{t,x^{n-1},c,\xi}^{2n+m}, R_{t,x^n,c,\xi}^{2n+m}\right\}$.

3.2. *The first stage of the cascade*

The flattening diffeomorphisms θ_1 transforms

1) the set of the fiber metabundles from (30) into the set of following ones

$$\left\{\Gamma_{x_i}^1\left(t,y_1,x_2,\ldots,x_n;c_1,c_i,\xi\right),\pi_{x_i}^1\left(c_1,\xi\right),R^1\times\Xi,\mathrm{G}_{\hat{c},\xi}^{x_i}\left(t,y_1,x_2,\ldots,x_n\right)^1\right\}_{i=1}^{n} \tag{32}$$

with the total spaces being

$$\left\{\begin{array}{l}\Gamma_{x_1}^1\left(t,y_1,x_2,\ldots,x_n;c_1,c_i,\xi\right)=\left\{y_1=0\right\},\\ \left\{\Gamma_{x_i}^1\left(t,y_1,x_2,\ldots,x_n;c_1,c_i,\xi\right)=\left(\begin{array}{l}g_i\left(t,y_1+\varphi_1\left(t,x_2,\ldots,x_n;c_1,\xi\right),x_2,\ldots,x_n,\xi\right)=\\ =g_i^1\left(t,y_1,x_2,\ldots,x_n;c_1,\xi\right)=c_i\end{array}\right)\right\}_{i=2}^{n}\end{array}\right\}. \tag{33}$$

Here we assume that

$$g_1\left(t,y_1+\varphi_1\left(t,x_2,\ldots,x_n;c_1,\xi\right),x_2,\ldots,x_n,\xi\right)-c_1\equiv 0\Big|_{y_1=0}; \tag{34}$$

2) the differential inclusion from (30) into the differential inclusion

$$\left\{\begin{array}{l}\dfrac{dy_1}{dt}\in\bigcup\limits_{\xi\in\Xi}f_1^1\left(t,y_1,x_2,\ldots,x_n;c_1,\xi\right),\dfrac{dx_2}{dt}\in\bigcup\limits_{\xi\in\Xi}f_2^1\left(t,y_1,x_2,\ldots,x_n;c_1,\xi\right),\ldots\\ \ldots,\dfrac{dx_n}{dt}\in\bigcup\limits_{\xi\in\Xi}f_n^1\left(t,y_1,x_2,\ldots,x_n;c_1,\xi\right)\end{array}\right\}, \tag{35}$$

where $f_1^1\left(t,y_1,x_2,\ldots,x_n;c_1,\xi\right)=f_1\left(t,y_1+\varphi_1\left(t,x_2,\ldots,x_n;c_1,\xi\right),x_2,\ldots,x_n,\xi\right)-\dfrac{\partial\varphi_1\left(t,x_2,\ldots,x_n;c_1,\xi\right)}{\partial t}-$ $-\sum\limits_{i=2}^{n}\dfrac{\partial\varphi_1\left(t,x_2,\ldots,x_n;c_1,\xi\right)}{\partial x_i}f_i\left(t,y_1+\varphi_1\left(t,x_2,\ldots,x_n;c_1,\xi\right),x_2,\ldots,x_n,\xi\right), f_i^1\left(t,y_1,x_2,\ldots,x_n;c_1,\xi\right)=$ $=f_i\left(t,y_1+\varphi_1\left(t,x_2,\ldots,x_n;c_1,\xi\right),x_2,\ldots,x_n,\xi\right), i=\left(2,\ldots,n\right)$.

The X_{t_0} transforms into $X_{t_0}^{y_1}=\theta_1\left(X_{t_0}\right)$ that is the orthogonal projection of X_{t_0} onto $R_{t,x^1,c,\xi}^{2n+m}$, that is $X_{t_0}^{y_1}\subseteq R_{t,x^1,c,\xi}^{2n+m}$.

Thus at the end of the first stage of the cascade of flattening diffeomorphisms we receive the following whole after-first-stage-of-cascade-of-diffeomorphisms system

$$\left\{ \begin{array}{l} \left\{ \Gamma^1_{x_i}\left(t,y_1,x_2,\dots,x_n;c_1,c_i,\xi\right), \pi^1_{x_i}\left(c_i,\xi\right), R^1\times\Xi, \mathrm{G}^{x_i}_{\hat{c}_i,\hat{\xi}}\left(t,y_1,x_2,\dots,x_n\right)^1 \right\}_{i=1}^{n}; \\ \left\{ \begin{array}{l} \dfrac{dy_1}{dt}\in\bigcup\limits_{\xi\in\Xi} f_1^1\left(t,y_1,x_2,\dots,x_n;c_1,\xi\right), \dfrac{dx_2}{dt}\in\bigcup\limits_{\xi\in\Xi} f_2^1\left(t,y_1,x_2,\dots,x_n;c_1,\xi\right),\dots \\ \dots,\dfrac{dx_n}{dt}\in\bigcup\limits_{\xi\in\Xi} f_n^1\left(t,y_1,x_2,\dots,x_n;c_1,\xi\right) \end{array} \right\} \end{array} \right\}, \tag{36}$$

where $f_1^1\left(t,y_1,x_2,\dots,x_n;c_1,\xi\right)\Big|_{\forall\left(t,y_1,x_2,\dots,x_n;c,\xi\right)\in\Gamma^1_{x_1}\left(t,y_1,x_2,\dots,x_n;c_1,\xi\right)=\{y_1=0\}}\equiv 0$, $y_{1,0}=0$.

3.3. *The second stage of the cascade*

The flattening diffeomorphisms θ_2 transforms

1) the set of the fiber metabundles from (36) into the set of following ones

$$\left\{ \Gamma^2_{x_i}\left(t,y_1,y_2,x_3,\dots,x_n;c_1,c_2,c_i,\xi\right), \pi^2_{x_i}\left(c_1,c_2,\xi\right), R^1\times\Xi, \mathrm{G}^{x_i}_{\hat{c},\hat{\xi}}\left(t,y_1,y_2,x_3,\dots,x_n\right)^2 \right\}_{i=1}^{n} \tag{37}$$

with the total spaces being

$$\left\{ \begin{array}{l} \Gamma^2_{x_1}\left(t,y_1,y_2,x_3,\dots,x_n;c_1,c_2,\xi\right)=\{y_1=0\}, \Gamma^2_{x_2}\left(t,y_1,y_2,x_3,\dots,x_n;c_1,c_2,\xi\right)=\{y_2=0\}, \\ \left\{ \begin{array}{l} \Gamma^2_{x_i}\left(t,y_1,x_2,\dots,x_n;c_1,c_2,c_i,\xi\right)= \\ =\left(\begin{array}{l} g_i^1\left(t,y_1,y_2+\varphi_2^1\left(t,y_1,x_3,\dots,x_n;c_1,c_2,\xi\right),x_3,\dots,x_n;c_1,\xi\right)= \\ =g_i^2\left(t,y_1,y_2,x_3,\dots,x_n;c_1,c_2,\xi\right)=c_i \end{array} \right) \end{array} \right\}_{i=3}^{n} \end{array} \right\}. \tag{38}$$

Here we assume that

$$g_2^1\left(t,y_1,y_2+\varphi_2^1\left(t,y_1,x_3,\dots,x_n;c_1,c_2,\xi\right),x_3,\dots,x_n,\xi\right)-c_2\equiv 0\Big|_{y_2=0}; \tag{39}$$

2) the differential inclusion from (36) into the differential inclusion

$$\left\{ \begin{array}{l} \dfrac{dy_1}{dt}\in\bigcup\limits_{\xi\in\Xi} f_1^2\left(t,y_1,y_2,x_3,\dots,x_n;c_1,c_2,\xi\right), \dfrac{dy_2}{dt}\in\bigcup\limits_{\xi\in\Xi} f_2^2\left(t,y_1,y_2,x_3,\dots,x_n;c_1,c_2,\xi\right), \\ \dfrac{dx_3}{dt}\in\bigcup\limits_{\xi\in\Xi} f_3^2\left(t,y_1,y_2,x_3,\dots,x_n;c_1,c_2,\xi\right),\dots,\dfrac{dx_n}{dt}\in\bigcup\limits_{\xi\in\Xi} f_n^2\left(t,y_1,y_2,x_3,\dots,x_n;c_1,c_2,\xi\right) \end{array} \right\}, \tag{40}$$

where $f_1^2\left(t,y_1,y_2,x_3,\dots,x_n;c_1,c_2,\xi\right)=f_1^1\left(t,y_1,y_2+\varphi_2^1\left(t,y_1,x_3,\dots,x_n;c_1,c_2,\xi\right),x_3,\dots,x_n;c_1,\xi\right)$,

$$f_2^2(t,y_1,y_2,x_3,...,x_n;c_1,c_2,\xi)=f_2^1\left(t,y_1,y_2+\varphi_2^1(t,y_1,x_3,...,x_n;c_1,c_2,\xi),x_3,...,x_n;c_1,\xi\right)-$$
$$-\frac{\partial\varphi_2^1(t,y_1,x_3,...,x_n;c_1,c_2,\xi)}{\partial t}-\frac{\partial\varphi_2^1(t,y_1,x_3,...,x_n;c_1,c_2,\xi)}{\partial y_1}f_1^2(t,y_1,y_2,x_3,...,x_n;c_1,c_2,\xi)-$$
$$-\sum_{i=3}^{n}\frac{\partial\varphi_2^1(t,y_1,x_3,...,x_n;c_1,c_2,\xi)}{\partial x_i}f_i^2(t,y_1,y_2,x_3,...,x_n;c_1,c_2,\xi),f_3^2(t,y_1,y_2,x_3,...,x_n;c_1,c_2,\xi)=$$
$$=f_3^1\left(t,y_1,y_2+\varphi_2^1(t,y_1,x_3,...,x_n;c_1,c_2,\xi),x_3,...,x_n;c_1,\xi\right),...,f_n^2(t,y_1,y_2,x_3,...,x_n;c_1,c_2,\xi)=$$
$$=f_n^1\left(t,y_1,y_2+\varphi_2^1(t,y_1,x_3,...,x_n;c_1,c_2,\xi),x_3,...,x_n;c_1,\xi\right).$$

The $X_{t_0}^{y_1}$ transforms into $X_{t_0}^{y_1y_2}=\theta_2\left(X_{t_0}^{y_1}\right)$ that is the orthogonal projection of $X_{t_0}^{y_1}$ onto $R_{t,y_1,x_3,...,x_n,c,\xi}^{2n+m}$, that is $X_{t_0}^{y_1y_2}\subseteq R_{t,y_1,x_3,...,x_n,c,\xi}^{2n+m}$.

Thus at the end of the second stage of the cascade of flattening diffeomorphisms we receive the following whole after-second-stage-of-cascade-of-diffeomorphisms system

$$\left\{\begin{array}{l}\left\{\Gamma_{x_i}^2(t,y_1,y_2,x_3,...,x_n;c_1,c_2,c_i,\xi),\pi_{x_i}^2(c_1,c_2,\xi),R^1\times\Xi,G_{\hat{c},\hat{\xi}}^{x_i}(t,y_1,y_2,x_3,...,x_n)^2\right\}_{i=1}^{n};\\ \left\{\begin{array}{l}\frac{dy_1}{dt}\in\bigcup_{\xi\in\Xi}f_1^2(t,y_1,y_2,x_3,...,x_n;c_1,c_2,\xi),\frac{dy_2}{dt}\in\bigcup_{\xi\in\Xi}f_2^2(t,y_1,y_2,x_3,...,x_n;c_1,c_2,\xi),\\ \frac{dx_3}{dt}\in\bigcup_{\xi\in\Xi}f_3^2(t,y_1,y_2,x_3,...,x_n;c_1,c_2,\xi),...,\frac{dx_n}{dt}\in\bigcup_{\xi\in\Xi}f_n^2(t,y_1,y_2,x_3,...,x_n;c_1,c_2,\xi)\end{array}\right\}\end{array}\right\}, \tag{41}$$

where $f_1^2(t,y_1,y_2,x_3,...,x_n;c_1,c_2,\xi)\Big|_{\forall(t,y_1,y_2,x_3,...,x_n;c,\xi)\in\Gamma_{x_1}^2(t,y_1,y_2,x_3,...,x_n;c_1,c_2,c_i,\xi)=\{y_1=0\}}\equiv 0, y_{1,0}=0$;

$f_2^2(t,y_1,y_2,x_3,...,x_n;c_1,c_2,\xi)\Big|_{\forall(t,y_1,y_2,x_3,...,x_n;c,\xi)\in\Gamma_{x_1}^2(t,y_1,y_2,x_3,...,x_n;c_1,c_2,c_i,\xi)=\{y_2=0\}}\equiv 0, y_{2,0}=0$.

3.4. *The third stage of the cascade*

The flattening diffeomorphisms θ_3 transforms

1) the set of the fiber metabundles from (41) into the set of following ones

$$\left\{\Gamma_{x_i}^3(t,y_1,y_2,y_3,x_4,...,x_n;c_1,c_2,c_3,c_i,\xi),\pi_{x_i}^3(c_1,c_2,c_3,\xi),R^1\times\Xi,G_{\hat{c},\hat{\xi}}^{x_i}(t,y_1,y_2,y_3,x_4,...,x_n)^3\right\}_{i=1}^{n} \tag{42}$$

with the total spaces being

$$\left\{\begin{array}{l}\Gamma_{x_1}^3\left(t, y_1, y_2, y_3, x_4, \ldots, x_n; c_1, c_2, c_3, \xi\right)=\left\{y_1=0\right\}, \\ \Gamma_{x_2}^3\left(t, y_1, y_2, y_3, x_4, \ldots, x_n; c_1, c_2, c_3, \xi\right)=\left\{y_2=0\right\}, \\ \Gamma_{x_3}^3\left(t, y_1, y_2, y_3, x_4, \ldots, x_n; c_1, c_2, c_3, \xi\right)=\left\{y_3=0\right\}, \\ \left\{\begin{array}{l}\Gamma_{x_i}^3\left(t, y_1, y_2, y_3, x_4, \ldots, x_n; c_1, c_2, c_3, c_i, \xi\right)= \\ =\left(\begin{array}{l}g_i^2\left(t, y_1, y_2, y_3+\varphi_3^2\left(t, y_1, y_2, x_4, \ldots, x_n; c_1, c_2, c_3, \xi\right), x_4, \ldots, x_n; c_1, c_{2,} \xi\right)= \\ =g_i^3\left(t, y_1, y_2, y_3, x_4, \ldots, x_n; c_1, c_2, c_3, \xi\right)=c_i\end{array}\right)\end{array}\right\}_{i=4}^n\end{array}\right\}. \quad (43)$$

Here again we assume that

$$g_3^2\left(t, y_1, y_2, y_3+\varphi_3^2\left(t, y_1, y_2, x_4, \ldots, x_n; c_1, c_2, c_3, \xi\right), x_4, \ldots, x_n; c_1, c_{2,} \xi\right)-c_3 \equiv 0\Big|_{y_3=0}; \quad (44)$$

2) the differential inclusion from (41) into the differential inclusion

$$\left\{\begin{array}{l}\dfrac{dy_1}{dt} \in \bigcup\limits_{\xi \in \Xi} f_1^3\left(t, y_1, y_2, y_3, x_4, \ldots, x_n; c_1, c_2, c_3, \xi\right), \\ \dfrac{dy_2}{dt} \in \bigcup\limits_{\xi \in \Xi} f_2^3\left(t, y_1, y_2, y_3, x_4, \ldots, x_n; c_1, c_2, c_3, \xi\right), \\ \dfrac{dy_3}{dt} \in \bigcup\limits_{\xi \in \Xi} f_3^3\left(t, y_1, y_2, y_3, x_4, \ldots, x_n; c_1, c_2, c_3, \xi\right), \\ \dfrac{dx_4}{dt} \in \bigcup\limits_{\xi \in \Xi} f_4^3\left(t, y_1, y_2, y_3, x_4, \ldots, x_n; c_1, c_2, c_3, \xi\right), \\ \ldots\ldots\ldots\ldots\ldots\ldots\ldots\ldots\ldots\ldots \\ \dfrac{dx_n}{dt} \in \bigcup\limits_{\xi \in \Xi} f_n^3\left(t, y_1, y_2, y_3, x_4, \ldots, x_n; c_1, c_2, c_3, \xi\right)\end{array}\right\}, \quad (45)$$

where

$$\begin{array}{l}f_1^3\left(t, y_1, y_2, y_3, x_4, \ldots, x_n; c_1, c_2, c_3, \xi\right)= \\ =f_1^2\left(t, y_1, y_2, y_3+\varphi_3^2\left(t, y_1, y_2, x_4, \ldots, x_n; c_1, c_2, c_3, \xi\right), x_4, \ldots, x_n; c_1, c_{2,} \xi\right), \\ f_2^3\left(t, y_1, y_2, y_3, x_4, \ldots, x_n; c_1, c_2, c_3, \xi\right)= \\ =f_2^2\left(t, y_1, y_2, y_3+\varphi_3^2\left(t, y_1, y_2, x_4, \ldots, x_n; c_1, c_2, c_3, \xi\right), x_4, \ldots, x_n; c_1, c_{2,} \xi\right),\end{array}$$

$$
\begin{aligned}
&f_3^3\left(t,y_1,y_2,y_3,x_4,\dots,x_n;c_1,c_2,c_3,\xi\right)=\\
&=f_3^2\left(t,y_1,y_2,y_3+\varphi_3^2\left(t,y_1,y_2,x_4,\dots,x_n;c_1,c_2,c_3,\xi\right),x_4,\dots,x_n;c_1,c_2,\xi\right)-\\
&-\frac{\partial\varphi_3^2\left(t,y_1,y_2,x_4,\dots,x_n;c_1,c_2,c_3,\xi\right)}{\partial t}-\frac{\partial\varphi_3^2\left(t,y_1,y_2,x_4,\dots,x_n;c_1,c_2,c_3,\xi\right)}{\partial y_1}\times\\
&\times f_1^3\left(t,y_1,y_2,y_3,x_4,\dots,x_n;c_1,c_2,c_3,\xi\right)-\frac{\partial\varphi_3^2\left(t,y_1,y_2,x_4,\dots,x_n;c_1,c_2,c_3,\xi\right)}{\partial y_2}\times\\
&\times f_2^3\left(t,y_1,y_2,y_3,x_4,\dots,x_n;c_1,c_2,c_3,\xi\right)-\sum_{i=4}^{n}\frac{\partial\varphi_3^2\left(t,y_1,y_2,x_4,\dots,x_n;c_1,c_2,c_3,\xi\right)}{\partial x_i}\times\\
&\times f_i^3\left(t,y_1,y_2,y_3,x_4,\dots,x_n;c_1,c_2,c_3,\xi\right),f_4^3\left(t,y_1,y_2,y_3,x_4,\dots,x_n;c_1,c_2,c_3,\xi\right)=\\
&=f_4^2\left(t,y_1,y_2,y_3+\varphi_3^2\left(t,y_1,y_2,x_4,\dots,x_n;c_1,c_2,c_3,\xi\right),x_4,\dots,x_n;c_1,c_2,\xi\right),\dots,\\
&f_n^3\left(t,y_1,y_2,y_3,x_4,\dots,x_n;c_1,c_2,c_3,\xi\right)=\\
&=f_n^3\left(t,y_1,y_2,y_3+\varphi_3^2\left(t,y_1,y_2,x_4,\dots,x_n;c_1,c_2,c_3,\xi\right),x_4,\dots,x_n;c_1,c_2,\xi\right).
\end{aligned}
$$

The $X_{t_0}^{y_1y_2}$ transforms into $X_{t_0}^{y_1y_2y_3}=\theta_3\left(X_{t_0}^{y_1y_2}\right)$ that is the orthogonal projection of $X_{t_0}^{y_1y_2}$ onto $R^{2n+m}_{t,y_1,y_2,x_4,\dots,x_n,c,\xi}$, that is $X_{t_0}^{y_1y_2y_3}\subseteq R^{2n+m}_{t,y_1,y_2,x_4,\dots,x_n,c,\xi}$.

Thus at the end of the third stage of the cascade of flattening diffeomorphisms we receive the following whole after-third-stage-of-cascade-of-diffeomorphisms system

$$
\left\{
\begin{array}{l}
\left\{\begin{array}{l}\Gamma_{x_i}^3\left(t,y_1,y_2,y_3,x_4,\dots,x_n;c_1,c_2,c_3,c_i,\xi\right),\pi_{x_i}^3\left(c_1,c_2,c_3,\xi\right),R^1\times\Xi,\\ G_{\hat{c},\hat{\xi}}^{x_i}\left(t,y_1,y_2,y_3,x_4,\dots,x_n\right)^3\end{array}\right\}_{i=1}^{n};\\
\left\{\begin{array}{l}
\dfrac{dy_1}{dt}\in\bigcup\limits_{\xi\in\Xi}f_1^3\left(t,y_1,y_2,y_3,x_4,\dots,x_n;c_1,c_2,c_3,\xi\right),\\
\dfrac{dy_2}{dt}\in\bigcup\limits_{\xi\in\Xi}f_2^3\left(t,y_1,y_2,y_3,x_4,\dots,x_n;c_1,c_2,c_3,\xi\right),\\
\dfrac{dy_3}{dt}\in\bigcup\limits_{\xi\in\Xi}f_3^3\left(t,y_1,y_2,y_3,x_4,\dots,x_n;c_1,c_2,c_3,\xi\right),\\
\dfrac{dx_4}{dt}\in\bigcup\limits_{\xi\in\Xi}f_4^3\left(t,y_1,y_2,y_3,x_4,\dots,x_n;c_1,c_2,c_3,\xi\right),\\
\dots\dots\dots\dots\dots\dots\dots\dots\dots\dots\dots\dots\\
\dfrac{dx_n}{dt}\in\bigcup\limits_{\xi\in\Xi}f_n^3\left(t,y_1,y_2,y_3,x_4,\dots,x_n;c_1,c_2,c_3,\xi\right)
\end{array}\right\}
\end{array}
\right\}, \qquad (46)
$$

where

$$f_1^3\left(t,y_1,y_2,y_3,x_4,\ldots,x_n;c_1,c_2,c_3,\xi\right)\Big|_{\forall(t,y_1,y_2,y_3,x_4,\ldots,x_n;c,\xi)\in\Gamma_{x_1}^3(t,y_1,y_2,y_3,x_3,\ldots,x_n;c_1,c_2,c_3,c_i,\xi)=\{y_1=0\}}\equiv 0,\, y_{1,0}=0;$$

$$f_2^3\left(t,y_1,y_2,y_3,x_4,\ldots,x_n;c_1,c_2,c_3,\xi\right)\Big|_{\forall(t,y_1,y_2,y_3,x_4,\ldots,x_n;c,\xi)\in\Gamma_{x_2}^3(t,y_1,y_2,y_3,x_3,\ldots,x_n;c_1,c_2,c_3,c_i,\xi)=\{y_2=0\}}\equiv 0,\, y_{2,0}=0;$$

$$f_3^3\left(t,y_1,y_2,y_3,x_4,\ldots,x_n;c_1,c_2,c_3,\xi\right)\Big|_{\forall(t,y_1,y_2,y_3,x_4,\ldots,x_n;c,\xi)\in\Gamma_{x_3}^3(t,y_1,y_2,y_3,x_3,\ldots,x_n;c_1,c_2,c_3,c_i,\xi)=\{y_3=0\}}\equiv 0,\, y_{3,0}=0.$$

3.5. The (k)-th stage of the cascade

From the whole after-$(k-1)$th-stage-of-cascade-of-diffeomorphisms system the flattening diffeomorphisms θ_k transforms

1) the set of its fiber metabundles into the set of following ones

$$\left\{\begin{matrix}\Gamma_{x_i}^k\left(t,y_1,\ldots,y_k,x_{k+1},\ldots,x_n;c_1,\ldots,c_k,c_i,\xi\right),\pi_{x_i}^k\left(c_1,\ldots,c_k,\xi\right),R^1\times\Xi,\\ \mathrm{G}_{\hat{c},\hat{\xi}}^{x_i}\left(t,y_1,\ldots,y_k,x_{k+1},\ldots,x_n\right)^k\end{matrix}\right\}_{i=1}^n \tag{47}$$

with the total spaces being

$$\left\{\begin{array}{l}\Gamma_{x_1}^k\left(t,y_1,\ldots,y_k,x_{k+1},\ldots,x_n;c_1,\ldots,c_k,\xi\right)=\{y_1=0\},\\ \ldots\ldots\ldots\ldots\ldots\ldots\ldots\ldots\ldots\ldots\ldots\ldots\\ \Gamma_{x_k}^k\left(t,y_1,\ldots,y_k,x_{k+1},\ldots,x_n;c_1,\ldots,c_k,\xi\right)=\{y_k=0\},\\ \left\{\begin{array}{l}\Gamma_{x_i}^k\left(t,y_1,\ldots,y_k,x_{k+1},\ldots,x_n;c_1,\ldots,c_k,c_i,\xi\right)=\\ =\left(\begin{array}{l}g_i^{k-1}(t,y_1,\ldots,y_{k-1},y_k+\varphi_k^{k-1}\left(t,y_1,\ldots,y_{k-1},x_{k+1},\ldots,x_n;c_1,\ldots,c_k,\xi\right),x_{k+1},\ldots,\\ x_n;c_1,\ldots,c_{k-1,}\xi)=g_i^k\left(t,y_1,\ldots,y_k,x_{k+1},\ldots,x_n;c_1,\ldots,c_k,\xi\right)=c_i\end{array}\right)\end{array}\right\}_{i=k+1}^n\end{array}\right\}. \tag{48}$$

Here we assume that

$$\left.\begin{array}{l}g_k^{k-1}\left(t,y_1,\ldots,y_{k-1},y_k+\varphi_k^{k-1}\left(t,y_1,\ldots,y_{k-1},x_{k+1},\ldots,x_n;c_1,\ldots,c_k,\xi\right),x_{k+1},\ldots,x_n;c_1,\ldots,c_{k-1,}\xi\right)-\\ -c_k\equiv 0\end{array}\right|_{y_k=0}; \tag{49}$$

2) its differential inclusion into the differential inclusion

$$\left\{\begin{array}{l} \frac{dy_1}{dt} \in \bigcup_{\xi\in\Xi} f_1^k\left(t, y_1, \ldots, y_k, x_{k+1}, \ldots, x_n; c_1, \ldots, c_k, \xi\right), \\ \ldots\ldots\ldots\ldots\ldots\ldots\ldots\ldots\ldots\ldots\ldots\ldots \\ \frac{dy_3}{dt} \in \bigcup_{\xi\in\Xi} f_k^k\left(t, y_1, \ldots, y_k, x_{k+1}, \ldots, x_n; c_1, \ldots, c_k, \xi\right), \\ \frac{dx_4}{dt} \in \bigcup_{\xi\in\Xi} f_{k+1}^k\left(t, y_1, \ldots, y_k, x_{k+1}, \ldots, x_n; c_1, \ldots, c_k, \xi\right), \\ \ldots\ldots\ldots\ldots\ldots\ldots\ldots\ldots\ldots\ldots\ldots\ldots \\ \frac{dx_n}{dt} \in \bigcup_{\xi\in\Xi} f_n^k\left(t, y_1, \ldots, y_k, x_{k+1}, \ldots, x_n; c_1, \ldots, c_k, \xi\right) \end{array}\right\}, \tag{50}$$

where

$$\begin{array}{l} f_1^k\left(t, y_1, \ldots, y_k, x_{k+1}, \ldots, x_n; c_1, \ldots, c_k, \xi\right) = \\ = f_1^{k-1}\left(t, y_1, \ldots, y_{k-1}, y_k + \varphi_k^{k-1}\left(t, y_1, \ldots, y_{k-1}, x_{k+1}, \ldots, x_n; c_1, \ldots, c_k, \xi\right), x_{k+1}, \ldots, x_n; c_1, \ldots, c_{k-1,} \xi\right), \ldots, \\ f_{k-1}^k\left(t, y_1, \ldots, y_k, x_{k+1}, \ldots, x_n; c_1, \ldots, c_k, \xi\right) = \\ = f_{k-1}^{k-1}\left(t, y_1, \ldots, y_{k-1}, y_k + \varphi_k^{k-1}\left(t, y_1, \ldots, y_{k-1}, x_{k+1}, \ldots, x_n; c_1, \ldots, c_k, \xi\right), x_{k+1}, \ldots, x_n; c_1, \ldots, c_{k-1,} \xi\right), \\ f_k^k\left(t, y_1, \ldots, y_k, x_{k+1}, \ldots, x_n; c_1, \ldots, c_k, \xi\right) = \\ = f_k^{k-1}\left(t, y_1, \ldots, y_{k-1}, y_k + \varphi_k^{k-1}\left(t, y_1, \ldots, y_{k-1}, x_{k+1}, \ldots, x_n; c_1, \ldots, c_k, \xi\right), x_{k+1}, \ldots, x_n; c_1, \ldots, c_{k-1,} \xi\right) - \\ - \frac{\partial \varphi_k^{k-1}\left(t, y_1, \ldots, y_{k-1}, x_{k+1}, \ldots, x_n; c_1, \ldots, c_k, \xi\right)}{\partial t} - \sum_{j=1}^{k-1} \frac{\partial \varphi_k^{k-1}\left(t, y_1, \ldots, y_{k-1}, x_{k+1}, \ldots, x_n; c_1, \ldots, c_k, \xi\right)}{\partial y_j} \times \\ \times f_j^k\left(t, y_1, \ldots, y_k, x_{k+1}, \ldots, x_n; c_1, \ldots, c_k, \xi\right) - \sum_{i=k+1}^{n} \frac{\partial \varphi_k^{k-1}\left(t, y_1, \ldots, y_{k-1}, x_{k+1}, \ldots, x_n; c_1, \ldots, c_k, \xi\right)}{\partial x_i} \times \\ \times f_i^k\left(t, y_1, \ldots, y_k, x_{k+1}, \ldots, x_n; c_1, \ldots, c_k, \xi\right), f_{k+1}^k\left(t, y_1, \ldots, y_k, x_{k+1}, \ldots, x_n; c_1, \ldots, c_k, \xi\right) = \\ = f_{k+1}^{k-1}\left(t, y_1, \ldots, y_{k-1}, y_k + \varphi_k^{k-1}\left(t, y_1, \ldots, y_{k-1}, x_{k+1}, \ldots, x_n; c_1, \ldots, c_k, \xi\right), x_{k+1}, \ldots, x_n; c_1, \ldots, c_{k-1,} \xi\right), \ldots, \\ f_n^k\left(t, y_1, \ldots, y_k, x_{k+1}, \ldots, x_n; c_1, \ldots, c_k, \xi\right) = \\ = f_n^{k-1}\left(t, y_1, \ldots, y_{k-1}, y_k + \varphi_k^{k-1}\left(t, y_1, \ldots, y_{k-1}, x_{k+1}, \ldots, x_n; c_1, \ldots, c_k, \xi\right), x_{k+1}, \ldots, x_n; c_1, \ldots, c_{k-1,} \xi\right). \end{array},$$

The $X_{t_0}^{y_1\cdots y_{k-1}}$ transforms into $X_{t_0}^{y_1\cdots y_k} = \theta_k\left(X_{t_0}^{y_1\cdots y_{k-1}}\right)$ that is the orthogonal projection of $X_{t_0}^{y_1\cdots y_{k-1}}$ onto $R^{2n+m}_{t,y_1,\ldots,y_{k-1},x_{k+1},\ldots,x_n,c,\xi}$, that is $X_{t_0}^{y_1\cdots y_k} \subseteq R^{2n+m}_{t,y_1,\ldots,y_{k-1},x_{k+1},\ldots,x_n,c,\xi}$.

Thus at the end of the second stage of the cascade of flattening diffeomorphisms we receive the following whole after-(k)th-stage-of-cascade-of-diffeomorphisms system

$$\left\{\begin{array}{l}\left\{\begin{array}{l}\Gamma_{x_i}^{k}\left(t,y_1,...,y_k,x_{k+1},...,x_n;c_1,...,c_k,c_i,\xi\right),\pi_{x_i}^{k}\left(c_1,...,c_3,\xi\right),R^1\times\Xi,\\ G_{\hat{c},\hat{\xi}}^{x_i}\left(t,y_1,...,y_k,x_{k+1},...,x_n\right)^k\end{array}\right\}_{i=1}^{n};\\ \left\{\begin{array}{l}\frac{dy_1}{dt}\in\bigcup_{\xi\in\Xi}f_1^k\left(t,y_1,...,y_k,x_{k+1},...,x_n;c_1,...,c_k,\xi\right),\\ \cdots\cdots\cdots\cdots\cdots\cdots\cdots\cdots\\ \frac{dy_3}{dt}\in\bigcup_{\xi\in\Xi}f_k^k\left(t,y_1,...,y_k,x_{k+1},...,x_n;c_1,...,c_k,\xi\right),\\ \frac{dx_4}{dt}\in\bigcup_{\xi\in\Xi}f_{k+1}^k\left(t,y_1,...,y_k,x_{k+1},...,x_n;c_1,...,c_k,\xi\right),\\ \cdots\cdots\cdots\cdots\cdots\cdots\cdots\cdots\\ \frac{dx_n}{dt}\in\bigcup_{\xi\in\Xi}f_n^k\left(t,y_1,...,y_k,x_{k+1},...,x_n;c_1,...,c_k,\xi\right)\end{array}\right\}\end{array}\right\}, \tag{51}$$

where

$$\begin{array}{l}f_1^k\left(t,y_1,...,y_k,x_{k+1},...,x_n;c_1,...,c_k,\xi\right)\Big|_{\forall(t,y_1,...,y_k,x_{k+1},...,x_n;c,\xi)\in\Gamma_{x_1}^k(t,y_1,...,y_k,x_{k+1},...,x_n;c_1,...,c_k,c_i,\xi)=\{y_1=0\}}\equiv 0,\ y_{1,0}=0\\ \cdots\cdots\cdots\cdots\cdots\cdots\cdots\cdots\cdots\cdots\cdots\cdots .\\ f_k^k\left(t,y_1,...,y_k,x_{k+1},...,x_n;c_1,...,c_k,\xi\right)\Big|_{\forall(t,y_1,...,y_k,x_{k+1},...,x_n;c,\xi)\in\Gamma_{x_k}^k(t,y_1,...,y_k,x_{k+1},...,x_n;c_1,...,c_k,c_i,\xi)=\{y_k=0\}}\equiv 0,\ y_{k,0}=0\end{array}$$

3.6. The (n)-th stage of the cascade

From the whole after-$(n-1)$th-stage-of-cascade-of-diffeomorphisms system the flattening diffeomorphisms θ_n transforms

1) the set of its fiber metabundles into the set of following ones

$$\left\{\Gamma_{x_i}^{n}\left(t,y_1,...,y_n;c_1,...,c_n,,\xi\right),\pi_{x_i}^{n}\left(c_1,...,c_n,\xi\right),R^1\times\Xi,G_{\hat{c},\hat{\xi}}^{x_i}\left(t,y_1,...,y_n\right)^n\right\}_{i=1}^{n} \tag{52}$$

with the total spaces being

$$\left\{\begin{array}{l}\Gamma_{x_1}^{n}\left(t,y_1,...,y_n;c_1,...,c_n,\xi\right)=\left\{y_1=0\right\},\\ \cdots\cdots\cdots\cdots\cdots\cdots\\ \Gamma_{x_n}^{n}\left(t,y_1,...,y_n;c_1,...,c_n,\xi\right)=\left\{y_n=0\right\}\end{array}\right\}. \tag{53}$$

Here we assume that

$$\left. \begin{array}{l} g_n^{n-1}\left(t,y_1,\dots,y_{n-1},y_n+\varphi_n^{n-1}\left(t,y_1,\dots,y_{n-1};c_1,\dots,c_n,\xi\right);c_1,\dots,c_{n-1,}\xi\right)- \\ -c_n\equiv 0 \end{array}\right|_{y_n=0} ; \tag{54}$$

2) its differential inclusion into the differential inclusion

$$\left\{ \begin{array}{l} \dfrac{dy_1}{dt}\in\bigcup\limits_{\xi\in\Xi} f_1^n\left(t,y_1,\dots,y_n;c,\xi\right), \\ \dots\dots\dots\dots\dots\dots\dots\dots\dots\dots\dots\dots \\ \dfrac{dx_n}{dt}\in\bigcup\limits_{\xi\in\Xi} f_n^n\left(t,y_1,\dots,y_n;c,\xi\right) \end{array}\right\} \Leftrightarrow \left\{\frac{dy}{dt}\in\bigcup_{\xi\in\Xi} f^n\left(t,y;c,\xi\right)\right\}, \tag{55}$$

where $f^n\left(t,y;c,\xi\right)=\left(f_1^n\left(t,y;c,\xi\right),\dots,f_n^n\left(t,y;c,\xi\right)\right), y=\left(y_1,\dots,y_n\right)$,

$$\left\{ \begin{array}{l} f_i^n\left(t,y_1,\dots,y_n,;c,\xi\right)= \\ =f_i^{n-1}\left(t,y_1,\dots,y_{n-1},y_n+\varphi_n^{n-1}\left(t,y_1,\dots,y_{n-1};c_1,\dots,c_n,\xi\right);c_1,\dots,c_{n-1,}\xi\right) \end{array}\right\}_{i=1}^{n-1},$$

$$f_n^n\left(t,y_1,\dots,y_n,;c,\xi\right)=f_n^{n-1}\left(t,y_1,\dots,y_{n-1},y_n+\varphi_n^{n-1}\left(t,y_1,\dots,y_{n-1};c_1,\dots,c_n,\xi\right);c_1,\dots,c_{n-1,}\xi\right)-$$

$$-\frac{\partial\varphi_n^{n-1}\left(t,y_1,\dots,y_{n-1};c_1,\dots,c_n,\xi\right)}{\partial t}-\sum_{j=1}^{n-1}\frac{\partial\varphi_n^{n-1}\left(t,y_1,\dots,y_{n-1};c_1,\dots,c_n,\xi\right)}{\partial y_j}f_j^n\left(t,y_1,\dots,y_n,;c,\xi\right).$$

The $X_{t_0}^{y_1\dots y_{n-1}}$ transforms into $X_{t_0}^{y_1\dots y_n}=\theta_k\left(X_{t_0}^{y_1\dots y_{n-1}}\right)$ that is the orthogonal projection of $X_{t_0}^{y_1\dots y_{n-1}}$ onto $R^{2n+m}_{t,y_1,\dots,y_{n-1},c,\xi}$ and $X_{t_0}^{y_1\dots y_n}\subseteq R^{2n+m}_{t,y_1,\dots,y_{n-1},c,\xi}$. Actually, after all the transformations we have obtained $X_{t_0}^{y_1\dots y_n}=0$.

If we fix the value of the vector of parameter $\xi=\hat{\xi}$ then the differential inclusion obtained after the (n)th stage of the cascade of flattening diffeomorphisms

$$\frac{dy}{dt}\in\bigcup_{\xi\in\Xi} f^n\left(t,y;c,\xi\right) \tag{56}$$

that we will call the canonical parametric differential inclusion, turns into the free dynamic system

$$\frac{dy}{dt}=f^n\left(t,y;c,\hat{\xi}\right), \tag{57}$$

which is the restriction of the differential inclusion (56) to $T \times R_y^n$. All its integral curves are identical and equal to the ray $T \subset R_t^1 \subset R_{y,t}^{n+1}$, namely

$$y_t\left(y_0,\hat{\xi}\right)=\left(t,y\left(t;y_0,\hat{\xi}\right)\right)\equiv\left(t,\left(\underbrace{y_1\left(t;y_0,\hat{\xi}\right)}_{\substack{\parallel\\0}},\ldots,\underbrace{y_n\left(t;y_0,\hat{\xi}\right)}_{\substack{\parallel\\0}}\right)\right)=\left(t,(0,\ldots,0)=\mathbf{0}\right)\forall t\in T. \quad (58)$$

Remark 4. *For the free dynamic system (57) all the phase components of all the initial points of its integral curves are equal to zero according to (58) that is*

$$y_{t_0}=\left(t=t_0,y_0=\left(y_{1,0},\ldots,y_{n,0}\right)=(0,\ldots,0)=\mathbf{0}\right). \quad (59)$$

The different values of these initial points are taken into consideration through the constant c *in the right-hand sides of the differential equations of (57) and the one-to-one correspondence between the initial point of any integral curve in* $T \times R_x^n$ *and the initial point of its counterpart in* $T \times R_y^n$ *is given by means of the relation (22), namely*

$$\begin{pmatrix}\left\{g_i\left(t_0,x_0;\xi\right)=c_i\right\}_{i=1}^n\\ \Updownarrow\\ g\left(t_0,x_0;\xi\right)=c\end{pmatrix}\Rightarrow x_0=\psi\left(t_0,c,\xi\right).$$

Thus at the end of the (n)th stage of the cascade of flattening diffeomorphisms we receive the following whole after-(n)th-stage-of-cascade-of-diffeomorphisms system

$$\left\{\begin{array}{l}\left\{\begin{array}{l}\Gamma_{x_i}^n\left(t,y_1,\ldots,y_n;c_1,\ldots,c_n,\xi\right),\\ \pi_{x_i}^n\left(c_1,\ldots,c_n,\xi\right),R^1\times\Xi,\\ \mathrm{G}_{\hat{c},\hat{\xi}}^{x_i}\left(t,y_1,\ldots,y_n\right)^n\end{array}\right\}_{i=1}^n;\\ \left\{\begin{array}{l}\dfrac{dy_1}{dt}\in\bigcup\limits_{\xi\in\Xi}f_1^n\left(t,y_1,\ldots,y_n;c,\xi\right),\\ \cdots\cdots\cdots\cdots\cdots\cdots\\ \dfrac{dx_n}{dt}\in\bigcup\limits_{\xi\in\Xi}f_n^n\left(t,y_1,\ldots,y_n;c,\xi\right)\end{array}\right\}\end{array}\right\}\Leftrightarrow\left\{\begin{array}{l}\left\{\Gamma_{x_i}^n\left(t,y;c,\xi\right),\pi_{x_i}^n\left(c,\xi\right),R^1\times\Xi,\mathrm{G}_{\hat{c},\hat{\xi}}^{x_i}\left(t,y\right)^n\right\}_{i=1}^n;\\ \left\{\dfrac{dy}{dt}\in\bigcup\limits_{\xi\in\Xi}f^n\left(t,y;c,\xi\right)\right\}\end{array}\right\}, \quad (60)$$

where $\left\{ f_i^n\left(t,y;c,\xi\right)\Big|_{\forall(t,y;c,\xi)\in\Gamma_{x_i}^n(t,y,c,\xi)=\{y_i=0\}} \equiv 0, y_{i,0}=0 \right\}_{i=1}^{n}$.

The cascade of sequential flattening diffeomorphisms θ creates the diffeomorphism $\overline{\varphi}$ that establishes the diffeomorphic correspondence between the initial system (30) and the ultimately-transformed one (60). Let us find it using the inverse cascade of sequential substitutions as follows

$$
\left\{
\begin{array}{l}
x_n = y_n + \varphi_n^{n-1}\left(t,y_1,...,y_{n-1};c_1,...,c_n,\xi\right) = \overline{\varphi}_n^{n-1}\left(t,y;c,\xi\right) \\
\downarrow \qquad\qquad\qquad\qquad x_n \\
\downarrow \text{-----------}\rightarrow \quad \downarrow_{x_n=\overline{\varphi}_n^{n-1}(\cdot)} \\
x_{n-1} = y_{n-1} + \varphi_{n-1}^{n-2}\left(t,y_1,...,y_{n-2},x_n;c_1,...,c_{n-1},\xi\right) = \\
\downarrow = y_{n-1} + \varphi_{n-1}^{n-2}\left(t,y_1,...,y_{n-2},\overline{\varphi}_n^{n-1}\left(t,y;c,\xi\right);c_1,...,c_{n-1},\xi\right) = \overline{\varphi}_{n-1}^{n-2}\left(t,y;c,\xi\right) \\
\downarrow \qquad\qquad\qquad\qquad x_{n-1} \qquad x_n \\
\downarrow \text{-----------}\rightarrow \quad \downarrow_{x_{n-1}==\overline{\varphi}_{n-1}^{n-2}(\cdot)} \; - \; \downarrow_{x_n==\overline{\varphi}_n^{n-1}(\cdot)} \\
x_{n-2} = y_{n-2} + \varphi_{n-2}^{n-3}\left(t,y_1,...,y_{n-3},x_{n-1},x_n;c_1,...,c_{n-2},\xi\right) = \\
\downarrow = y_{n-2} + \varphi_{n-2}^{n-3}\left(t,y_1,...,y_{n-3},\overline{\varphi}_{n-1}^{n-2}\left(t,y;c,\xi\right),\overline{\varphi}_n^{n-1}\left(t,y;c,\xi\right);c_1,...,c_{n-2},\xi\right) = \overline{\varphi}_{n-2}^{n-3}\left(t,y;c,\xi\right) \\
\downarrow \qquad\qquad\qquad\qquad x_{n-2} \qquad x_{n-1} \qquad x_n \\
\downarrow \text{-----------}\rightarrow \quad \downarrow_{x_n=\overline{\varphi}_{n-2}^{n-3}(\cdot)} \; -- \; \downarrow_{x_{n-1}=\overline{\varphi}_{n-1}^{n-2}(\cdot)} \; -- \; \downarrow_{x_n=\overline{\varphi}_n^{n-1}(\cdot)} \\
\cdots\cdots\cdots\cdots\cdots\cdots\cdots\cdots\cdots\cdots\cdots\cdots \\
\downarrow \qquad\qquad\qquad\qquad x_{k+1} \qquad x_n \\
\downarrow \text{-----------}\rightarrow \quad \downarrow_{x_n=\overline{\varphi}_{k+1}^{k}(\cdot)} \; -- \; \downarrow_{x_n=\overline{\varphi}_n^{n-1}(\cdot)} \\
x_k = y_k + \varphi_k^{k-1}\left(t,y_1,...,y_{k-1},x_{k+1},........,x_n;c_1,...,c_k,\xi\right) = \\
\downarrow \\
\downarrow = y_k + \varphi_k^{k-1}\left(t,y_1,...,y_{k-1},\overline{\varphi}_{k+1}^{k}\left(t,y;c,\xi\right),...,\overline{\varphi}_n^{n-1}\left(t,y;c,\xi\right);c_1,...,c_k,\xi\right) = \overline{\varphi}_k^{k-1}\left(t,y;c,\xi\right) \\
\downarrow \\
\downarrow \qquad\qquad\qquad\qquad x_k \qquad x_n \\
\downarrow \text{-----------}\rightarrow \quad \downarrow_{x_k=\overline{\varphi}_k^{k-1}(\cdot)} \; ---- \; \downarrow_{x_n=\overline{\varphi}_n^{n-1}(\cdot)} \\
\cdots\cdots\cdots\cdots\cdots\cdots\cdots\cdots\cdots\cdots\cdots\cdots \\
\downarrow \qquad x_2 \qquad x_n \\
\downarrow \text{----}\rightarrow \downarrow_{x_2=\overline{\varphi}_2^{1}(\cdot)} \; -- \; \downarrow_{x_n=\overline{\varphi}_n^{n-1}(\cdot)} \\
x_1 = y_1 + \varphi_1\left(t,x_2,.......,x_n;c_1,\xi\right) = \varphi_1\left(t,\overline{\varphi}_2^{1}\left(t,y;c,\xi\right),...,\overline{\varphi}_n^{n-1}\left(t,y;c,\xi\right);c_1,\xi\right) = \overline{\varphi}_1^{0}\left(t,y;c,\xi\right)
\end{array}
\right\} . \quad (61)
$$

Thus we receive

$$\overline{\varphi}=\left(\overline{\varphi}_1^0(t,y;c,\xi),\overline{\varphi}_2^1(t,y;c,\xi),\ldots,\overline{\varphi}_k^{k-1}(t,y;c,\xi),\ldots,\overline{\varphi}_{n-1}^{n-2}(t,y;c,\xi),\overline{\varphi}_n^{n-1}(t,y;c,\xi)\right); \tag{62}$$

$$\overline{\varphi}:\left\{\begin{array}{l}\left\{\begin{array}{l}\Gamma_{x_i}^n(t,y;c,\xi),\pi_{x_i}^n(c,\xi),\\ R^1\times\Xi,\mathrm{G}_{\hat{c},\hat{\xi}}^{x_i}(t,y)^n\end{array}\right\}_{i=1}^n;\\ \left\{\dfrac{dy}{dt}\in\bigcup\limits_{\xi\in\Xi}f^n(t,y;c,\xi),\right\}\end{array}\right\}\to\left\{\begin{array}{l}\left\{\begin{array}{l}\Gamma_{x_i}(t,x;c_i,\xi),\pi_{x_i}(c_i,\xi),\\ R^1\times\Xi,\mathrm{G}_{\hat{c}_i,\hat{\xi}}^{x_i}(t,x)\end{array}\right\}_{i=1}^n;\\ \dfrac{dx}{dt}\in\bigcup\limits_{\xi\in\Xi}f(t,x;\xi)\end{array}\right\} \tag{63}$$

or

$$\left\{\begin{array}{l}\left\{\begin{array}{l}\Gamma_{x_i}(t,x;c_i,\xi),\pi_{x_i}(c_i,\xi),\\ R^1\times\Xi,\mathrm{G}_{\hat{c}_i,\hat{\xi}}^{x_i}(t,x)\end{array}\right\}_{i=1}^n;\\ \dfrac{dx}{dt}\in\bigcup\limits_{\xi\in\Xi}f(t,x;\xi)\end{array}\right\}\xrightarrow{\theta:\ \theta_1\Rightarrow\ldots\Rightarrow\theta_k\Rightarrow\ldots\Rightarrow\theta_n}\left\{\begin{array}{l}\left\{\begin{array}{l}\Gamma_{x_i}^n(t,y;c,\xi),\pi_{x_i}^n(c,\xi),\\ R^1\times\Xi,\mathrm{G}_{\hat{c},\hat{\xi}}^{x_i}(t,y)^n\end{array}\right\}_{i=1}^n;\\ \left\{\dfrac{dy}{dt}\in\bigcup\limits_{\xi\in\Xi}f^n(t,y;c,\xi)\right\}\end{array}\right\}. \tag{64}$$

Definition 2. The system of the relations (60) is called the canonical form of the representation of the initial system (30) and the diffeomorphism $\overline{\varphi}$ establishing the diffeomorphic correspondence between the initial system (30) and its canonical form is called the canonizing diffeomorphism.

Fig. 4 illustrates the action of the cascade of sequential flattening diffeomorphisms θ and of the canonizing diffeomorphism $\overline{\varphi}$.

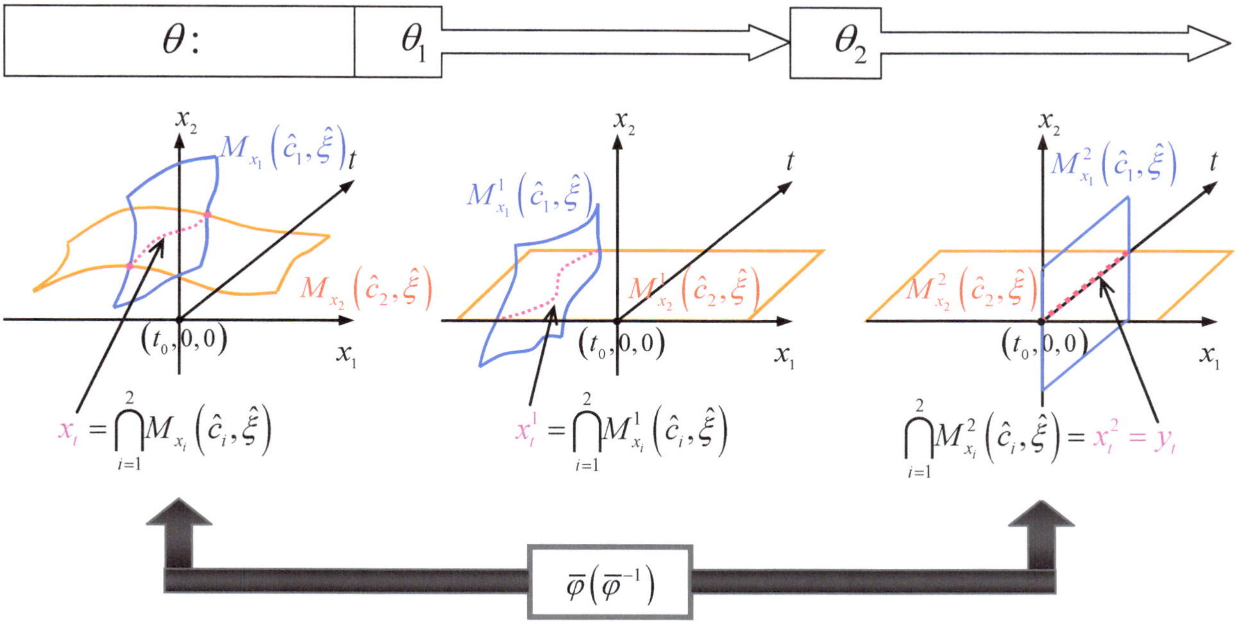

FIG. 4

Definition 3. We say that a given parametric function $\overline{\varphi}_i^{i-1}(t,y;c,\xi), i\in(1,...,n)$ of a real variable $t\in T$ is uniformly convergent in y to $\overline{\varphi}_i^{i-1}(t,0;c,\xi)$ if for any arbitrary small $\varepsilon>0$, given in advance, there exists a such neighborhood $Y(\varepsilon)$ of the point $\hat{y}=0$ that for any $y\in Y(\varepsilon)$ and $\forall(t,c,\xi)\in T\times R^n\times\Xi$ we have

$$\left\|\overline{\varphi}_i^{i-1}(t,y;c,\xi)-\overline{\varphi}_i^{i-1}(t,0;c,\xi)\right\|<\varepsilon, \tag{65}$$

where $y=(y_1,...,y_n)$ is considered a vector of parameter in this definition.

We will designate the uniform convergence of $\overline{\varphi}_i^{i-1}(t,y;c,\xi)$ as

$$\overline{\varphi}_i^{i-1}(t,y;c,\xi)\overbrace{---\rightarrow}^{y\rightarrow 0}\overline{\varphi}_i^{i-1}(t,0;c,\xi). \tag{66}$$

Assumption. All considered spaces and manifolds before and after applying diffeomorphisms intersect transversely.

This means that all spaces and manifolds are in general positions without any special cases of degeneration, topological pathology, tangency, etc.

Lemma 1. The geometrical-topological relation between the total spaces of the fiber metabundle (5) $\left\{\Gamma_x(t,x;x_0,\xi),\pi_x(x_0,\xi),X_{t_0}\times\Xi,\mathrm{G}_{\hat{x}_0,\hat{\xi}}(t,x)\right\}$ and the set of the fiber metabundles from (30) $\left\{\Gamma_{x_i}(t,x;\mathrm{c}_i,\xi),\pi_{x_i}(\mathrm{c}_i,\xi),R^1\times\Xi,\mathrm{G}^{x_i}_{\hat{c}_i,\hat{\xi}}(t,x)\right\}_{i=1}^n$ is described with the following diagrams

1. For the total spaces

$$\begin{array}{ccc} & \Gamma_x(t,x;x_0,\xi)=\bigcap_{i=1}^n\Gamma_{x_i}(t,x;\mathrm{c}_i,\xi) & \\ \nwarrow\!\searrow & & \swarrow\!\nearrow \\ & \psi(\psi^{-1}):R^n\times\Xi\rightleftarrows X_{t_0} & \end{array}\;; \tag{67}$$

2. For the typical fibers

$$\begin{array}{ccc} & \Gamma_x(t,x;\hat{x}_0,\hat{\xi})=\bigcap_{i=1}^n\Gamma_{x_i}(t,x;\hat{\mathrm{c}}_i,\hat{\xi})=x_t(\hat{x}_0,\hat{\xi}) & \\ \nwarrow\!\searrow & & \swarrow\!\nearrow \\ & \psi(\psi^{-1}):(\hat{\mathrm{c}}_i,\hat{\xi})\rightleftarrows\hat{x}_0 & \end{array}\;. \tag{68}$$

Deciphering analytically the diagram for the typical fibers of the fiber metabundles we have

$$x_t\left(\hat{x}_0,\hat{\xi}\right)=\left\{g\left(t,x;\hat{\xi}\right)=\hat{c};\forall\left(t,x\right)\in T\times R_x^n,\forall\left(\hat{c},\xi\right)\in R^n\times\Xi\right\}, \tag{69}$$

where

$$\hat{x}_0=\psi\left(t_0,\hat{c},\hat{\xi}\right). \tag{70}$$

In other words a specific integral curve $x_t\left(\hat{x}_0,\hat{\xi}\right)$ is the intersection of n of one-codimensional invariant manifolds $\left\{M_{x_i}\left(\hat{c}_i,\hat{\xi}\right)\right\}_{i=1}^n$ that are the level sections of the first integrals $g\left(t,x;\hat{\xi}\right)=c$ with the hyperplanes $c=\hat{c}$. The correspondence between the initial point $x_{t_0}\left(\hat{x}_0,\hat{\xi}\right)=\left(t_0,\hat{x}_0\right)$ of the integral curve and the real numbers $\hat{c}$ is established by the equation (70).

Proposition. The canonizing diffeomorphism $\overline{\varphi}\left(\overline{\varphi}^{-1}\right)$, conserving the intersections and their transversality, has the following properties

$$\begin{gathered}\overline{\varphi}^{-1}\left(\bigcap_{i=1}^{n}\Gamma_{x_i}^{n}\left(t,y;c,\xi\right)\right)\equiv\left|\bigcap_{i=1}^{n}\Gamma_{x_i}^{n}\left(t,y;c,\xi\right)=\left\{y_i=0\right\}_{i=1}^{n}\right|=\\ =\bigcap_{i=1}^{n}\overline{\varphi}^{-1}\left(\Gamma_{x_i}^{n}\left(t,y;c,\xi\right)\right)=\bigcap_{i=1}^{n}\Gamma_{x_i}\left(t,x;c_i,\xi\right)=\Gamma_x\left(t,x;x_0,\xi\right)\\ \searrow \qquad\qquad \nearrow\\ \psi\left(\psi^{-1}\right):R^n\times\Xi\rightleftarrows X_{t_0}\end{gathered} \tag{71}$$

and

$$\begin{gathered}\overline{\varphi}^{-1}\left(\bigcap_{i=1}^{n}\Gamma_{x_i}^{n}\left(t,y;\hat{c},\hat{\xi}\right)\right)\equiv\left|\bigcap_{i=1}^{n}\Gamma_{x_i}^{n}\left(t,y;\hat{c},\hat{\xi}\right)=\left\{y_i=0\right\}_{i=1}^{n}=y_t\left(0,\hat{\xi}\right)\right|=\\ =\bigcap_{i=1}^{n}\overline{\varphi}^{-1}\left(\Gamma_{x_i}^{n}\left(t,y;\hat{c},\hat{\xi}\right)\right)=\bigcap_{i=1}^{n}\Gamma_{x_i}\left(t,x;\hat{c}_i,\hat{\xi}\right)=\Gamma_x\left(t,x;\hat{x}_0,\hat{\xi}\right)=x_t\left(\hat{x}_0,\hat{\xi}\right).\\ \searrow \qquad\qquad \nearrow\\ \psi\left(\psi^{-1}\right):\left(\hat{c}_i,\hat{\xi}\right)\rightleftarrows\hat{x}_0\end{gathered} \tag{72}$$

Consider the set of the coverings $\left\{\left(p_i\left(c,\xi\right),T\times R_y^n\times R^n\times\Xi,S_i\right)\right\}_{i=1}^n$, consisting of the covering spaces $\left\{S_i\right\}_{i=1}^n$, the base spaces $\left\{T\times R_y^n\times R^n\times\Xi\right\}_{i=1}^n$ and the covering mappings

$\left\{\left(p_i(c,\xi):T\times R_y^n\times R^n\times\Xi\to S_i\right)\right\}_{i=1}^n$, which is generated by the equations of the differential inclusion of the total system written in the canonical form (60), namely

$$\left\{p_i(c,\xi):T\times R_y^n\times R^n\times\Xi\to S_i=\left\{s_i=f_i^n(t,y;c,\xi);(t,y)\in T\times R_y^n,(c,\xi)\in R^n\times\Xi\right\}\right\}_{i=1}^n,\quad(73)$$

where $\left\{f_i^n(t,y;c,\xi)\Big|_{\forall(t,y;c,\xi)\in\Gamma_{x_i}^n(t,y,c,\xi)=\{y_i=0\}}\equiv 0, y_{i,0}=0\right\}_{i=1}^n$.

Definition 4. The mapping $p(c,\xi)=\left(p_1(c,\xi),...,p_n(c,\xi)\right)$ is called the total covering mapping of the collection of all the coverings $\left\{\left(p_i(c,\xi),T\times R_y^n\times R^n\times\Xi,S_i\right)\right\}_{i=1}^n$.

Theorem 1. Consider the initial system (30) and its canonical representation (60) obtained by means of the canonizing diffeomorphism $\bar{\varphi}$. Let

1. $p(\hat{c},\hat{\xi})$ being the restriction of total covering mapping $p(c,\xi)$ to $T\times R_y^n$ belongs to the class of $\mathbf{A}$ – mappings, that is $p(\hat{c},\hat{\xi})\in[p]^{\mathbf{A}}$ [5, p. 66], [6, p. 18];
2. $\bar{\varphi}^{-1}(t,y;\hat{c},\hat{\xi})$ being the restriction of the canonizing diffeomorphism $\bar{\varphi}^{-1}(t,y;c,\xi)$ to $T\times R_y^n$ is uniformly convergent to $\bar{\varphi}^{-1}(t,0;\hat{c},\hat{\xi})$ in y. Suppose this condition is true for all $\hat{c}\in R^n,\hat{\xi}\in\Xi$, that is

$$\bar{\varphi}^{-1}(t,y;\hat{c},\hat{\xi})\overset{y\to 0}{\overbrace{---\to}}\bar{\varphi}^{-1}(t,0;\hat{c},\hat{\xi})\forall\left(\hat{c}\in R^n,\hat{\xi}\in\Xi\right).\quad(74)$$

Then the canonical form of the definite-positive auxiliary functions $W(y)$ of Lyapunov functions $V(t,y)$ for the canonical parametric differential inclusion (60)

$$W(y)=\sum_{i=1}^n y_i^2\quad(75)$$

ensures

a) the asymptotical stability of the integral curve $y_t(\hat{y}_0,\hat{\xi})=\left(t,y(t;\hat{y}_0,\hat{\xi})\right)\equiv(t,\mathbf{0})\forall t\in T$ starting from the initial point $(t=t_0,\hat{y}_0=\mathbf{0})$ of the free dynamic system

$$\frac{dy}{dt} = f^n\left(t, y; \hat{c}, \hat{\xi}\right), \tag{76}$$

which is the restriction of the canonical differential inclusion (60)

$$\frac{dy}{dt} \in \bigcup_{\xi \in \Xi} f^n\left(t, y; c, \xi\right)$$

to $T \times R_y^n$;

b) the asymptotical stability of the integral curves $x_t\left(\hat{x}_0, \hat{\xi}\right)$ starting from the initial point $\hat{x}_0 \in X_{t_0}$ of the free dynamical system (2), namely

$$\frac{dx}{dt} = f\left(t, x; \hat{\xi}\right),$$

which is the restriction of the initial differential inclusion (1) to $T \times R_y^n$, where $\hat{x}_0 = \psi\left(t_0, \hat{c}, \hat{\xi}\right)$ (see (22)).

The Theorem 1 is formulated according to the Lyapunov's second method for stability as it is expounded in [7]. The proof of this theorem and all the ones later on in the monograph can be easily conducted on the basis of the methodology and techniques developed in [5]. However we will prove it using Lyapunov functions in order to keep to the consistency of the analysis.

Proof. Construct the definite-positive Lyapunov function on the basis of the auxiliary function $W(y)$ shown in (75) in the following form

$$V(t, y) = W(y)\left(\lambda + e^{-t}\right), \tag{77}$$

where $\lambda \geq 1$.

Its total derivative equal to 0 if $y = \mathbf{0}$

$$\frac{dV(t, y)}{dt} = 2\left(\lambda + e^{-t}\right)\left\{\sum_{i=1}^{n} y_i \frac{dy_i}{dt}\right\} - e^{-t}\sum_{i=1}^{n} y_i^2$$

is a definite-negative function since $y_i \dfrac{dy_i}{dt} < 0 \forall i \in (1, \ldots, n)$ due to the property of $\mathbf{A}-$mappings.

It means that integral curve $y_t\left(\hat{y}_0, \hat{\xi}\right) = \left(t, y\left(t; \hat{y}_0, \hat{\xi}\right)\right) \equiv (t, \mathbf{0}) \forall t \in T$ of the system (76) is asymptotically stable.

According to (71) we have the smooth bijective map

$$x = \overline{\varphi}^{-1}\left(t, y; c, \hat{\xi}\right), \tag{78}$$

which links all corresponding integral curves $x_t\left(\hat{x}_0, \hat{\xi}\right)$ and $y_t\left(\hat{y}_0, \hat{\xi}\right)$ of the systems (2) and (76) respectively in pairs. In the terms of algebraic topology, the relation (78) also introduces the homotopy of the integral curves $x_t\left(\hat{x}_0, \hat{\xi}\right)$ of the initial system (2) with y as a vector of homotopic parameters executing their smooth deformation and translation. We define the distance between two any different integral curves of the above-mentioned systems in the following way

$$\left\| x_t\left(\hat{x}_0', \hat{\xi}\right), x_t\left(\hat{x}_0, \hat{\xi}\right) \right\| = \sqrt{\sum_{i=1}^{n} \left(x_i\left(t; \hat{x}_0', \hat{\xi}\right) - x_i\left(t; \hat{x}_0, \hat{\xi}\right) \right)^2},$$

$$\left\| y_t\left(\hat{y}_0', \hat{\xi}\right), y_t\left(\hat{y}_0, \hat{\xi}\right) \right\| = \sqrt{\sum_{i=1}^{n} \left(y_i\left(t; \hat{y}_0', \hat{\xi}\right) - y_i\left(t; \hat{y}_0, \hat{\xi}\right) \right)^2}.$$

The asymptotical stability of $y_t\left(\hat{y}_0, \hat{\xi}\right) = \left(t, y\left(t; \hat{y}_0, \hat{\xi}\right)\right) \equiv (t, \mathbf{0})$ means that for any arbitrary small $\delta > 0$, given in advance, there exists such a point of time $t_\delta > t_0$ that

$$\left\| y_t\left(\hat{y}_0, \hat{\xi}\right), y_t\left(\hat{y}_0', \hat{\xi}\right) \right\| = \sqrt{\sum_{i=1}^{n} \left[y_i\left(t; \hat{y}_0', \hat{\xi}\right) \right]^2} < \delta \forall t \ge t_\delta.$$

Combining the above-written inequality with (78), the definition of uniform convergence of parametric function (Definition 3) and the second condition of Theorem 1 we deduce that for any arbitrary small $\varepsilon > 0$, given in advance, such a point of time t_ε can be always found that

$$\left\| x_t\left(\hat{x}_0', \hat{\xi}\right), x_t\left(\hat{x}_0, \hat{\xi}\right) \right\| = \sqrt{\sum_{i=1}^{n} \left(x_i\left(t; \hat{x}_0', \hat{\xi}\right) - x_i\left(t; \hat{x}_0, \hat{\xi}\right) \right)^2} < \varepsilon \forall t \ge t_\varepsilon$$

and

$$\left\| y_t\left(\hat{y}_0, \hat{\xi}\right), y_t\left(\hat{y}_0', \hat{\xi}\right) \right\| = \sqrt{\sum_{i=1}^{n} \left[y_i\left(t; \hat{y}_0', \hat{\xi}\right) \right]^2} < \delta(\varepsilon) \forall t \ge t_\varepsilon.$$

This directly leads us to asymptotical stability of $x_t\left(\hat{x}_0, \hat{\xi}\right)$. The theorem is proved.

Corollary 1.1 The n-dimensional invariant manifolds $\left\{\Gamma_{x_i}\left(t,x;\hat{c}_i,\hat{\xi}\right)\right\}_{i=1}^{n}$ and the integral curve

$$x_t\left(\hat{x}_0,\hat{\xi}\right)=\Gamma_x\left(t,x;\hat{x}_0,\hat{\xi}\right)=\bigcap_{i=1}^{n}\Gamma_{x_i}\left(t,x;\hat{c}_i,\hat{\xi}\right)$$

are at least mosaic ω-attractors, that is the ones belong to the class of ω-attractors

$$x_t\left(\hat{x}_0,\hat{\xi}\right)=\Gamma_x\left(t,x;\hat{x}_0,\hat{\xi}\right)\in[A^{\omega}],\Gamma_{x_i}\left(t,x;\hat{c}_i,\hat{\xi}\right)\in[A^{\omega}].$$

In conclusion it is expedient to make the following important remark. It is obvious that there exists an infinite number of the functions, which satisfy the conditions of the basic theorem, namely

$$W(y)=\sum_{i=1}^{n}a_i y_i^l$$

where $a_i\in R^+, l=2\tilde{l}, \tilde{l}\in N^+$. From the expressions (75) and (77) it is clear why we call the former form of the function $W(y)$ canonical.

4. CLASSIFICATIONAL STABILITY OF TYPICAL FIBER OF FIBER SUBBUNDLES AND METABUNDLES

In this section we will consider the *classificational stability* of the typical fiber, which is a ω-attractor in the structure of fiber metabundle (25) through its two fiber subbundles (26) and (27). The concept of this stability has been introduced in [5, p. 96]. It will open us the possibilities to investigate the classificational stability of the total typical fiber, which is an asymptotically stable integral curve belonging to the class of ω-attractor, of the metabundle (5) through the fiber subbundles (6) and (7). In other words, we would like to see under what conditions the integral curve preserves its asymptotical stability when we fiddle with the vectors of parameters ξ and c or x_0.

The typical fiber of the fiber metabundle also as of the fiber subbundles (26) and (27) is a n-dimensional invariant manifold $\Gamma_{x_i}\left(t,x;\hat{x}_0,\hat{\xi}\right)\xleftrightarrow{\hat{c}_i=g_i\left(t_0,\hat{x}_0;\hat{\xi}\right)}M_{x_i}\left(\hat{c}_i,\hat{\xi}\right)$.

I. The topological and analytical criteria of the classificational stability of the typical fiber of the fiber subbundle (26), namely

$$\left\{\Gamma_{x_i}\left(t,x;\hat{x}_0,\xi\right),\pi_{x_i}\left(\hat{x}_0,\xi\right)\xleftrightarrow{\hat{c}_i=g_i\left(t_0,\hat{x}_0;\xi\right)}\pi_{x_i}\left(\hat{c}_i,\xi\right),\Xi,\mathrm{G}^{x_i}_{\hat{x}_0,\xi}\left(t,x\right)\right\}$$

can be formulated on the base of Theorems 3 and 8 of [6]. Further in the book we will consider the collection of the fiber bundles

$$\left\{\Gamma_{x_i}\left(t,x;\hat{x}_0,\xi\right),\pi_{x_i}\left(\hat{x}_0,\xi\right)\xleftrightarrow{\hat{c}_i=g_i\left(t_0,\hat{x}_0;\xi\right)}\pi_{x_i}\left(\hat{c}_i,\xi\right),\Xi,\mathrm{G}^{x_i}_{\hat{x}_0,\hat{\xi}}\left(t,x\right)\right\}_{i=1}^{n}$$

and the collections of their total subspaces $\left\{\Gamma_{x_i}\left(t,x;\hat{x}_0,\xi\right)\right\}_{i=1}^{n}$ and typical fibers

$$\left\{\Gamma_{x_i}\left(t,x;\hat{x}_0,\hat{\xi}\right)\xleftrightarrow{\hat{c}_i=g_i\left(t_0,\hat{x}_0;\hat{\xi}\right)}M_{x_i}\left(\hat{c}_i,\hat{\xi}\right)\right\}_{i=1}^{n}.$$

Theorem 2. Let there exists a m-dimensional neighborhood $\Xi_{\hat{\xi}}\subset\Xi$ of the point $\hat{\xi}$ such that the following conditions hold true for all $\xi\in\Xi_{\hat{\xi}}$:

1. Topological criterion:

$$p_i\left(\hat{c},\xi\right)\in\left[p_i\right]^{\mathbf{A}} \tag{79}$$

2. Analytical criterion:

 2.1. The classificational rank $\bar{s}$ for the covering mapping $p_i(\hat{c},\xi)$ preserves the odd parity.

 2.2. The negative sign of the partial derivative $\left.\left(\frac{\partial^{\bar{s}} f_i^n(t,y;\hat{c},\xi)}{\partial y_i^{\bar{s}}}\right)\right|_{y_i\equiv 0}$ is preserved, that is

$$sign\left.\left(\frac{\partial^{\bar{s}} f_i^n(t,y;\hat{c},\xi)}{\partial y_i^{\bar{s}}}\right)\right|_{y_i\equiv 0}=sign\left.\left(\frac{\partial^{\bar{s}} f_i^n(t,y;\hat{c},\hat{\xi})}{\partial y_i^{\bar{s}}}\right)\right|_{y_i\equiv 0}. \tag{80}$$

Then $\left\{\Gamma_{x_i}(t,x;\hat{x}_0,\hat{\xi})\xleftrightarrow{\hat{c}_i=g_i(t_0,\hat{x}_0;\hat{\xi})}M_{x_i}(\hat{c}_i,\hat{\xi})\right\}\in[\mathrm{A}^{\omega}]$ is classificationally stable at the point $\hat{\xi}$ meaning that

$$\left\{\Gamma_{x_i}(t,x;\hat{x}_0,\xi)\xleftrightarrow{\hat{c}_i=g_i(t_0,\hat{x}_0;\xi)}M_{x_i}(\hat{c}_i,\xi)\right\}\in[\mathrm{A}^{\omega}]\forall\xi\in\Xi_{\hat{\xi}} \tag{81}$$

Corollary 2.1 If at the point $\xi=\hat{\xi}\in\Xi$ the classificational rank $\bar{s}$ equals the unit and the sign of the partial derivative $\left.\left(\frac{\partial^{\bar{s}} f_i^n(t,y;\hat{c},\xi)}{\partial y_i^{\bar{s}}}\right)\right|_{y_i\equiv 0}$ is negative, that is

$$sign\left.\left(\frac{\partial^{\bar{s}} f_i^n(t,y;\hat{c},\hat{\xi})}{\partial y_i^{\bar{s}}}\right)\right|_{y_i\equiv 0}=-1,\bar{s}=1, \tag{82}$$

then $\left\{\Gamma_{x_i}(t,x;\hat{x}_0,\hat{\xi})\xleftrightarrow{\hat{c}_i=g_i(t_0,\hat{x}_0;\hat{\xi})}M_{x_i}(\hat{c}_i,\hat{\xi})\right\}\in[\mathrm{A}^{\omega}]$ is always classificationally stable and there is no necessity to satisfy the two above-mentioned preservation laws (2.1) and (2.2) for all points $\xi\in\Xi_{\hat{\xi}}$.

It is known that

$$\Gamma_{\hat{x}_0}(t,x;\hat{x}_0,\hat{\xi})=\left\{x_t(\hat{x}_0,\hat{\xi})\in X_t(\hat{\xi})\right\}=\bigcap_{i=1}^{n}\Gamma_{x_i}(t,x;\hat{x}_0,\hat{\xi}), \tag{83}$$

where $\Gamma_{\hat{x}_0}(t,x;\hat{x}_0,\hat{\xi})$ is the typical fiber being an integral curve of the fiber subbundle (6).

Corollary 2.2 Let all n typical fibers $\left\{\Gamma_{x_i}\left(t,x;\hat{x}_0,\hat{\xi}\right)\xleftrightarrow{\hat{c}_i=g_i\left(t_0,\hat{x}_0;\hat{\xi}\right)}M_{x_i}\left(\hat{c}_i,\hat{\xi}\right)\right\}\in[\mathrm{A}^{\omega}]$ of the collection $\left\{\Gamma_{x_i}\left(t,x;\hat{x}_0,\hat{\xi}\right)\xleftrightarrow{\hat{c}_i=g_i\left(t_0,\hat{x}_0;\hat{\xi}\right)}M_{x_i}\left(\hat{c}_i,\hat{\xi}\right)\right\}_{i=1}^{n}$ are classificationally stable at the point $\hat{\xi}\in\Xi_{\hat{\xi}}$, then the typical fiber (83) being the integral curve $x_t\left(\hat{x}_0,\hat{\xi}\right)$ asymptotically stable at the point $\hat{\xi}$ is also classificationally stable at this point.

In geometrical terms, all integral curves $x_t\left(\hat{x}_0,\xi\right)$ composing the bouquet $\bigcup_{\xi\in\Xi_{\hat{\xi}}} x_t\left(\hat{x}_0,\xi\right)$, each of which corresponds to its own point $\xi\in\Xi_{\hat{\xi}}$ and comes out from the common point $x_{t_0}=\left(t_0,\hat{x}_0\right)$, are asymptotically stable. In topological terms, the property of asymptotical stability is preserved for the restriction of the total subspace of the fiber subbundle (6) $\Gamma_x\left(t,x;\hat{x}_0,\xi\right)=\left\{x_t\left(\hat{x}_0,\xi\right)=\left(t,x\left(t;\hat{x}_0,\xi\right)\right)\forall\xi\in\Xi\Leftrightarrow\bigcup_{\forall\hat{\xi}\in\Xi}x_t\left(\hat{x}_0,\hat{\xi}\right),\left(t,x\right)\in T\times R_x^n;\hat{x}_0\in X_{t_0}\right\}$ to $\Xi_{\hat{\xi}}\subset\Xi$.

Definition 5. If the conditions of Corollary 2.2 holds true for $\forall\xi\in\Xi$ then the asymptotically stable integral curve $x_t\left(\hat{x}_0,\hat{\xi}\right)$ is called classificationally stable in the entire parameter manifold Ξ.

Definition 6. If the asymptotically stable integral curves $x_t\left(x_0,\hat{\xi}\right)$ are classificationally stable in the entire parameter manifold Ξ and $\forall x_0\in X_{t_0}\xleftrightarrow{c=g\left(t_0,x_0;\xi\right)\Leftrightarrow x_0=\psi\left(t_0,c,\xi\right)}\forall c\in R^n$ then the parametric differential inclusion (1) is called the one having the global asymptotical stability.

II. Now consider the classificational stability of the typical fiber

$$\Gamma_{x_i}\left(t,x;\hat{x}_0,\hat{\xi}\right)\xleftrightarrow{\hat{c}_i=g_i\left(t_0,\hat{x}_0;\hat{\xi}\right)}M_{x_i}\left(\hat{c}_i,\hat{\xi}\right)=$$

$$=\left\{g_i\left(t,x;\hat{\xi}\right)=\hat{c}_i\Big|_{\hat{c}_i=g_i\left(t_0,\hat{x}_0;\hat{\xi}\right)}\Leftrightarrow x_i=\varphi_i\left(t,x^i;t_0,\hat{x}_0,\hat{\xi}\right),\left(t,x\right)\in T\times R_x^n;\hat{c}_i\in R^1,\hat{\xi}\in\Xi\right\}$$

in the structure of the fiber subbundle (27) $\left\{\Gamma_{x_i}\left(t,x;x_0,\hat{\xi}\right),\pi_{x_i}\left(c_i,\hat{\xi}\right),R^1,\mathrm{G}_{\hat{x}_0,\hat{\xi}}^{x_i}\left(t,x\right)\right\}$ with the total subspace

$$\Gamma_{x_i}\left(t,x;x_0,\hat{\xi}\right)=F_{x_i}\left(\hat{\xi}\right)=$$
$$=\left\{\bigcup_{\forall c_i\in R^1}\left\{g_i\left(t,x;\hat{\xi}\right)=c_i\Big|_{c_i=g_i\left(t_0,x_0;\hat{\xi}\right)}\Leftrightarrow x_i=\varphi_i\left(t,x^i;t_0,x_0,\hat{\xi}\right)\right\},(t,x)\in T\times R_x^n;\hat{\xi}\in\Xi\right\}$$

and the base subspace R^1.

In order to formulate the conditions of classificational stability in this case we need to turn our attention to the quotient space $X_{t_0}/L_{x_i}\left(\hat{c}_i,\xi\right)$, which elements are $(n-1)$-dimensional submanifolds

$$L_{x_i}\left(\hat{c}_i,\hat{\xi}\right)=\left\{\hat{c}_i=g_i\left(t_0,x_0;\hat{\xi}\right)\Rightarrow x_{i,0}=\tilde{\varphi}_i\left(t_0,x_0^i,\hat{\xi};\hat{c}_i\right),x_0\in X_{t_0};\hat{\xi}\in\Xi\right\}$$

of the n-dimensional invariant manifold $\Gamma_{x_i}\left(t,x;\hat{x}_0,\hat{\xi}\right)\xleftrightarrow{\hat{c}_i=g_i\left(t_0,\hat{x}_0;\hat{\xi}\right)}M_{x_i}\left(\hat{c}_i,\hat{\xi}\right)$. The former ones serve the latter ones as the initial points x_{t_0} do the integral curves x_t , that is they are the initial elements from where $\Gamma_{x_i}\left(t,x;\hat{x}_0,\hat{\xi}\right)\xleftrightarrow{\hat{c}_i=g_i\left(t_0,\hat{x}_0;\hat{\xi}\right)}M_{x_i}\left(\hat{c}_i,\hat{\xi}\right)$ start. The quotient space $X_{t_0}/L_{x_i}\left(\hat{c}_i,\hat{\xi}\right)$ can be considered a $(n-1)$-dimensional foliation with $L_{x_i}\left(c_i,\hat{\xi}\right)$ as leaves, where $c_i\in R^1$.

Let us introduce in $X_{t_0}/L_{x_i}\left(\hat{c}_i,\hat{\xi}\right)$ the distance between two arbitrary leaves $L_{x_i}\left(\hat{c}_i'',\hat{\xi}\right)$ and $L_{x_i}\left(\hat{c}_i',\hat{\xi}\right)$ as follows

$$\left\|L_{x_i}\left(\hat{c}_i'',\hat{\xi}\right),L_{x_i}\left(\hat{c}_i',\hat{\xi}\right)\right\|=\inf_{\forall x_0^i\in pr_\perp\left(X_{t_0}\to R_{x^i}^{n-1}\right)}\left|\tilde{\varphi}_i\left(t_0,x_0^i,\hat{\xi};\hat{c}_i''\right)-\tilde{\varphi}_i\left(t_0,x_0^i,\hat{\xi};\hat{c}_i'\right)\right|, \tag{84}$$

where $pr_\perp\left(X_{t_0}\to R_{x^i}^{n-1}\right)$ is the orthogonal projection of the set X_{t_0} to the space $R_{x^i}^{n-1}$.

It is obvious that this distance can be also defined in the following way

$$\left\|L_{x_i}\left(\hat{c}_i'',\hat{\xi}\right),L_{x_i}\left(\hat{c}_i',\hat{\xi}\right)\right\|=\left|\hat{c}_i''-\hat{c}_i'\right| . \tag{85}$$

They are two equivalent definitions of the distance between two arbitrary leaves and we can use either of them.

Definition 7. The typical fiber $\Gamma_{x_i}\left(t,x;\hat{x}_0,\hat{\xi}\right)\xleftrightarrow{\hat{c}_i=g_i\left(t_0,\hat{x}_0;\hat{\xi}\right)\Rightarrow\hat{x}_{i,0}=\tilde{\varphi}_i\left(t_0,\hat{x}_0^i,\hat{\xi};\hat{c}_i\right)}M_{x_i}\left(\hat{c}_i,\hat{\xi}\right)\in[\mathrm{A}^\omega]$ is called classificationally stable in the structure of the fiber subbundle (27)

$\left\{\Gamma_{x_i}\left(t,x;x_0,\hat{\xi}\right),\pi_{x_i}\left(c_i,\hat{\xi}\right),R^1,\ G^{x_i}_{\hat{x}_0,\hat{\xi}}\left(t,x\right)\right\}$ if there exists a one-dimensional neighborhood, $C_{\hat{c}_i}\subset R^1$, of the point $\hat{c}$ such that for all $c_i\in C_{\hat{c}_i}$ the following condition holds true

$$\Gamma_{x_i}\left(t,x;x_0,\hat{\xi}\right)\xleftrightarrow{c_i=g_i\left(t_0,x_0;\hat{\xi}\right)\Rightarrow x_{i,0}=\tilde{\varphi}_i\left(t_0,x_0^i,\hat{\xi};c_i\right)}M_{x_i}\left(c_i,\hat{\xi}\right)\in[\mathrm{A}^\omega]. \tag{86}$$

This definition is based on how we have defined the distance between two arbitrary leaves in (85). If we would like to use the relation (84) then we have to replace the final part of the Definition 7 with "…if there exists a n-dimensional neighborhood, $EL_{x_i}\left(\hat{c}_i,\hat{\xi}\right)\subset X_{t_0}$, of the a $(n-1)$-dimensional leaf $L_{x_i}\left(\hat{c}_i,\hat{\xi}\right)$ such that for all $L_{x_i}\left(c_i,\hat{\xi}\right)\in EL_{x_i}\left(\hat{c}_i,\hat{\xi}\right)$ the following condition holds true…" If the condition of the classificational stability (86) is valid for all $c_i\in R^1$ or $EL_{x_i}\left(\hat{c}_i,\hat{\xi}\right)\equiv X_{t_0}$, then the total subspace $\Gamma_{x_i}\left(t,x;x_0,\hat{\xi}\right)$ has global classificational stability in the structure of the fiber subbundle (27).

According to Theorem 4 from [6] we have

Theorem 3. The typical fiber

$$\Gamma_{x_i}\left(t,x;\hat{x}_0,\hat{\xi}\right)\xleftrightarrow{\hat{c}_i=g_i\left(t_0,\hat{x}_0;\hat{\xi}\right)}M_{x_i}\left(\hat{c}_i,\hat{\xi}\right)=\left\{\begin{array}{l} g_i\left(t,x;\hat{\xi}\right)=\hat{c}_i\big|_{\hat{c}_i=g_i\left(t_0,\hat{x}_0;\hat{\xi}\right)}\Leftrightarrow \\ x_i=\varphi_i\left(t,x^i;t_0,\hat{x}_0,\hat{\xi}\right), \\ \left(t,x\right)\in T\times R^n_x;\hat{c}_i\in R^1,\hat{\xi}\in\Xi \end{array}\right\}\in[\mathrm{A}^\omega]$$

of the fiber subbundle (27) $\left\{\Gamma_{x_i}\left(t,x;x_0,\hat{\xi}\right),\pi_{x_i}\left(c_i,\hat{\xi}\right),R^1,G^{x_i}_{\hat{x}_0,\hat{\xi}}\left(t,x\right)\right\}$ is always classificationally stable at the point $\hat{x}_0\in X_{t_0}\xleftrightarrow{\hat{c}_i=g_i\left(t_0,\hat{x}_0;\hat{\xi}\right)\Leftrightarrow x_{i,0}=\tilde{\varphi}_i\left(t_0,x_0^i,\hat{\xi};\hat{c}_i\right)}\hat{c}_i\in R^1$ in contrast to ω-repellers and ω-shunts.

Theorem 4. Let

$$\Gamma_{\hat{x}_0}\left(t,x;\hat{x}_0,\hat{\xi}\right)=\left\{x_t\left(\hat{x}_0,\hat{\xi}\right)\in X_t\left(\hat{\xi}\right)\right\}=\bigcap_{i=1}^{n}\Gamma_{x_i}\left(t,x;\hat{x}_0,\hat{\xi}\right)\text{ and }\left\{\Gamma_{x_i}\left(t,x;\hat{x}_0,\hat{\xi}\right)\right\}_{i=1}^{n}\in[\mathrm{A}^\omega],$$

then the integral curve $x_t\left(\hat{x}_0,\hat{\xi}\right)$ is asymptotically stable.

There doesn't make sense to talk about the classificational stability of the integral curve $x_t\left(\hat{x}_0,\hat{\xi}\right)$ in this case because it automatically means its asymptotical stability.

Definition 8. If all the integral curves of the free dynamic system (2), that is $\forall \hat{x}_0 \in X_{t_0}$, are asymptotically stable, then the entire system (2) is called asymptotically stable.

Corollary 4.1. If all the total subspaces $\left\{\Gamma_{x_i}\left(t,x;x_0,\hat{\xi}\right)\right\}_{i=1}^{n} \in [\mathrm{A}^{\omega}]$ have global classificational stability in the structure of the corresponding fiber subbundles $\left\{\Gamma_{x_i}\left(t,x;x_0,\hat{\xi}\right),\pi_{x_i}\left(c_i,\hat{\xi}\right),R^1,\mathrm{G}^{x_i}_{\hat{x}_0,\hat{\xi}}\left(t,x\right)\right\}_{i=1}^{n}$, then free dynamic system (2)

$$\frac{dx}{dt}=f\left(t,x;\hat{\xi}\right)$$

is asymptotically stable.

Definition 9. If in the structure of the parametric differential inclusion (1)

$$\frac{dx}{dt}\in\bigcup_{\xi\in\Xi}f\left(t,x;\xi\right)$$

all the free dynamic system (2)

$$\frac{dx}{dt}=f\left(t,x;\xi\right),$$

that is $\forall\xi\in\Xi$, are asymptotical stable, then the entire inclusion (1) is asymptotical stable.

Corollary 4.2. If in the structure of the collection of the fiber metabundle (9)

$$\left\{\left\{\Gamma_x\left(t,x;x_0,\xi\right),\pi_x\left(\hat{x}_0,\xi\right),\Xi,\mathrm{G}_{x_0,\hat{\xi}}\left(t,x\right)\right\}\right\}_{i=1}^{n}$$

all their corresponding typical fibers $\left\{\Gamma_{x_i}\left(t,x;x_0,\hat{\xi}\right)\right\}_{i=1}^{n} \in [\mathrm{A}^{\omega}]$ have global classificational stability and this property is preserved by the section projection 2 $\pi_x\left(\hat{x}_0,\xi\right)$, then the entire parametric differential inclusion (1) is asymptotical stable.

Taking account of

$$\overline{\varphi}:\left\{\begin{array}{l}\left\{\begin{array}{l}\Gamma^{n}_{x_i}\left(t,y;c,\xi\right),\pi^{n}_{x_i}\left(c,\xi\right),\\ R^1\times\Xi,\mathrm{G}^{x_i}_{\hat{c},\hat{\xi}}\left(t,y\right)^{n}\end{array}\right\}_{i=1}^{n};\\ \left\{\dfrac{dy}{dt}\in\bigcup_{\xi\in\Xi}f^{n}\left(t,y;c,\xi\right),\right\}\end{array}\right\}\rightarrow\left\{\begin{array}{l}\left\{\begin{array}{l}\Gamma_{x_i}\left(t,x;\mathrm{c}_i,\xi\right),\pi_{x_i}\left(\mathrm{c}_i,\xi\right),\\ R^1\times\Xi,\mathrm{G}^{x_i}_{\hat{c}_i,\hat{\xi}}\left(t,x\right)\end{array}\right\}_{i=1}^{n};\\ \dfrac{dx}{dt}\in\bigcup_{\xi\in\Xi}f\left(t,x;\xi\right)\end{array}\right\},$$

Theorem 1 can be expanded over the entire parametric differential inclusion (1).

Theorem 5. Let

1. The total covering mapping $p(c,\xi)=(p_1(c,\xi),\dots,p_n(c,\xi))\in[p]^{A}$;
2. $\overline{\varphi}(t,y;c,\xi)\overbrace{---\to}^{y\to 0}\overline{\varphi}(t,0;c,\xi)\forall(c,\xi)\in R^n\times\Xi$.

Then the canonical form of the positive definite auxiliary functions $W(y)$ of Lyapunov functions $V(t,y)$ for the canonical parametric differential inclusion (56)

$$W(y)=\sum_{i=1}^{n} y_i^2$$

ensures

a) the asymptotical stability of the very (56)

$$\frac{dy}{dt}\in\bigcup_{\xi\in\Xi} f^n(t,y;c,\xi);$$

b) the asymptotical stability of the initial differential inclusion (1)

$$\frac{dx}{dt}\in\bigcup_{\xi\in\Xi} f(t,x;\xi).$$

Corollary 5.1. If under the conditions of Theorem 5 we will consider the restriction $p(c,\hat{\xi})$ of $p(c,\xi)$ to $T\times R_y^n\times R^n$, then the canonical form of the definite-positive auxiliary functions $W(y)$ of Lyapunov functions $V(t,y)$ for the canonical parametric differential inclusion (56)

$$W(y)=\sum_{i=1}^{n} y_i^2$$

insures the asymptotical stability just as the free dynamic system (57)

$$\frac{dy}{dt}=f^n(t,y;c,\hat{\xi})$$

so also the free dynamic system (2)

$$\frac{dx}{dt}=f(t,x;\hat{\xi}).$$

Example

Consider the following differential inclusion

$$\left\{\begin{aligned}\frac{dx_1}{dt} &= \frac{-\frac{d\eta(\bullet)}{dt}\left[2x_1x_2 - x_2^2\right] + \frac{d\eta(\bullet)}{dt}\gamma(\bullet) + \frac{d\gamma(\bullet)}{dt}\eta(\bullet)}{2\eta(\bullet)x_2} \\ \frac{dx_2}{dt} &= \frac{\frac{d\gamma(\bullet)}{dt}\eta(\bullet) - \frac{d\eta(\bullet)}{dt}\left[x_2^2 - \gamma(\bullet)\right]}{2\eta(\bullet)x_2}\end{aligned}\right\}, \tag{87}$$

where $\eta(\bullet) = \eta(t,\xi_1)$, $\gamma(\bullet) = \gamma(t,\xi_2,\xi_3,\xi_4)$, $\xi = (\xi_1,\xi_2,\xi_3,\xi_4) \in R^4$, $2\eta(\bullet)x_2 > 0 \forall t \in [0;+\infty[$.

The complete set of independent parametric first integrals is

$$\left\{\begin{aligned}&(x_1 - x_2)\eta(t,\xi_1) = c_1, \\ &\frac{x_2^2 - \gamma(t,\xi_2,\xi_3,\xi_4)}{x_1 - x_2} = c_2\end{aligned}\right\}. \tag{88}$$

The general solution of the initial differential inclusion (87) is given below

$$\left\{\begin{aligned}x_1 &= \frac{c_1}{\eta(t,\xi_1)} + \sqrt{\frac{c_1c_2}{\eta(t,\xi_1)} + \gamma(t,\xi_2,\xi_3,\xi_4)}, \\ x_2 &= \sqrt{\frac{c_1c_2}{\eta(t,\xi_1)} + \gamma(t,\xi_2,\xi_3,\xi_4)}\end{aligned}\right\}. \tag{89}$$

Thus, the whole initial system is

$$I_x = \left\{\begin{aligned}&\left\{\begin{aligned}&\Gamma_{x_1}(t,x_1,x_2;c_1,\xi) = \left\{g_1(t,x_1,x_2;\xi) = (x_1 - x_2)\eta(t,\xi_1) = c_1\right\}, \\ &\Gamma_{x_2}(t,x_1,x_2;c_2,\xi) = \left\{g_2(t,x_1,x_2;\xi) = \frac{x_2^2 - \gamma(t,\xi_2,\xi_3,\xi_4)}{x_1 - x_2} = c_2\right\}\end{aligned}\right\}; \\ &\left\{\begin{aligned}\frac{dx_1}{dt} &= \frac{-\frac{d\eta(\bullet)}{dt}\left[2x_1x_2 - x_2^2\right] + \frac{d\eta(\bullet)}{dt}\gamma(\bullet) + \frac{d\gamma(\bullet)}{dt}\eta(\bullet)}{2\eta(\bullet)x_2}, \\ \frac{dx_2}{dt} &= \frac{\frac{d\gamma(\bullet)}{dt}\eta(\bullet) - \frac{d\eta(\bullet)}{dt}\left[x_2^2 - x_2^2\gamma(\bullet)\right]}{2\eta(\bullet)x_2}\end{aligned}\right\}\end{aligned}\right\} \tag{90}$$

The cascade of flattening diffeomorphisms consists only of two stages:

$$\theta=\{\theta_1 \Rightarrow \theta_2\}=\left(\begin{array}{c} x_1 = y_1 + x_2 + \dfrac{c_1}{\eta(t,\xi_1)} \\ \downarrow \\ x_2 = y_2 + \sqrt{\gamma(t,\xi_2,\xi_3,\xi_4)+c_2\left[y_1+\dfrac{c_1}{\eta(t,\xi_1)}\right]} \end{array}\right). \tag{91}$$

The canonizing diffeomorphism

$$\bar{\varphi}^{-1}=\left\{\begin{array}{l} x_1 = y_1 + y_2 + \dfrac{c_1}{\eta(t,\xi_1)} + \sqrt{\gamma(t,\xi_2,\xi_3,\xi_4)+c_2\left[y_1+\dfrac{c_1}{\eta(t,\xi_1)}\right]} \\ x_2 = y_2 + \sqrt{\gamma(t,\xi_2,\xi_3,\xi_4)+c_2\left[y_1+\dfrac{c_1}{\eta(t,\xi_1)}\right]} \end{array}\right\} \tag{92}$$

transforms the initial differential inclusion (87) into

$$\left\{\begin{array}{l} \dfrac{dy_1}{dt}=-\dfrac{1}{\eta(\bullet)}\dfrac{d\eta(\bullet)}{dt}y_1=f_1^2(t,y_1,y_2;c_1,c_2,\xi), \\ \dfrac{dy_2}{dt}=\dfrac{\dfrac{d\gamma(\bullet)}{dt}\eta(\bullet)-\dfrac{d\eta(\bullet)}{dt}\left[\left(y_2+\sqrt{\gamma(\bullet)+c_2\left[y_1+\dfrac{c_1}{\eta(\bullet)}\right]}\right)^2-\gamma(\bullet)\right]}{2\eta(\bullet)\left(y_2+\sqrt{\gamma(\bullet)+c_2\left[y_1+\dfrac{c_1}{\eta(\bullet)}\right]}\right)}- \\ -\dfrac{1}{2}\dfrac{\dfrac{d\gamma(\bullet)}{dt}-\dfrac{c_2}{\eta(\bullet)}\dfrac{d\eta(\bullet)}{dt}\left[y_1+\dfrac{c_1}{\eta(\bullet)}\right]}{\sqrt{\gamma(\bullet)+c_2\left[y_1+\dfrac{c_1}{\eta(\bullet)}\right]}}=f_2^2(t,y_1,y_2;c_1,c_2,\xi) \end{array}\right\}. \tag{93}$$

We obtain the following canonical representation of the initial system I_x described by (90)

$$I_y = \left\{ \left\{ \begin{array}{l} \Gamma^2_{x_1}(\bullet) = \\ = \{y_1 = 0\}, \\ \Gamma^2_{x_2}(\bullet) = \\ = \{y_2 = 0\} \end{array} \right\}; \left\{ \begin{array}{l} \frac{dy_1}{dt} = -\frac{1}{\eta(\bullet)}\frac{d\eta(\bullet)}{dt} y_1 = f^2_1(t, y_1, y_2; c_1, c_2, \xi), \\ \frac{dy_2}{dt} = \frac{\frac{d\gamma(\bullet)}{dt}\eta(\bullet) - \frac{d\eta(\bullet)}{dt}\left[\left(y_2 + \sqrt{\gamma(\bullet) + c_2\left[y_1 + \frac{c_1}{\eta(\bullet)}\right]}\right)^2 - \gamma(\bullet)\right]}{2\eta(\bullet)\left(y_2 + \sqrt{\gamma(\bullet) + c_2\left[y_1 + \frac{c_1}{\eta(\bullet)}\right]}\right)} - \\ -\frac{1}{2}\frac{\frac{d\gamma(\bullet)}{dt} - \frac{c_2}{\eta(\bullet)}\frac{d\eta(\bullet)}{dt}\left[y_1 + \frac{c_1}{\eta(\bullet)}\right]}{\sqrt{\gamma(\bullet) + c_2\left[y_1 + \frac{c_1}{\eta(\bullet)}\right]}} = f^2_2(t, y_1, y_2; c_1, c_2, \xi) \end{array} \right\} \right\}, \quad (94)$$

where $\Gamma^2_{x_i}(\bullet) = \Gamma^2_{x_i}(t, y_1, y_2; c_i, \xi), i \in (1, 2)$.

Ultimately, we have

$$I_x \xleftrightarrow{\bar{\varphi}(\bar{\varphi}^{-1})} I_y . \quad (95)$$

To use analytical criteria of the classificational stability of the typical fibers of the fiber bundles $\Gamma^2_{x_1}(\bullet)$ and $\Gamma^2_{x_2}(\bullet)$, we need to derive $\frac{\partial f^2_1(t, y_1, y_2; c_1, c_2, \xi)}{\partial y_1}$ and $\frac{\partial f^2_2(t, y_1, y_2; c_1, c_2, \xi)}{\partial y_2}$. We have

$$\begin{array}{l} \frac{\partial f^2_1(t, y_1, y_2; c_1, c_2, \xi)}{\partial y_1} = -\frac{1}{\eta(\bullet)}\frac{d\eta(\bullet)}{dt}, \\ \frac{\partial f^2_2(t, y_1, y_2; c_1, c_2, \xi)}{\partial y_2} = -\frac{1}{2\eta(\bullet)W(\bullet)^2}\left\{\frac{d\eta(\bullet)}{dt}W(\bullet)^2 + \frac{d\gamma(\bullet)}{dt}\eta(\bullet) + \frac{d\eta(\bullet)}{dt}\gamma(\bullet)\right\}, \end{array} \quad (96)$$

where $W(\bullet) = y_2 + \sqrt{\gamma(\bullet) + c_2\left[y_1 + \frac{c_1}{\eta(\bullet)}\right]}$.

On the two-dimensional plane $\{y_1 = 0\} \in R^3_{t, y_1, y_2}$ and along the straight line being the axis t $\{y_1 = 0\} \cap \{y_2 = 0\} \in R^3_{t, y_1, y_2}$ the partial derivative $\frac{\partial f^2_1(t, y_1, y_2; c_1, c_2, \xi)}{\partial y_1}$ takes the same form, namely

$$\left.\frac{\partial f_1^2(t,y_1,y_2;c_1,c_2,\xi)}{\partial y_1}\right|_{y_1=0} = \left.\frac{\partial f_1^2(t,y_1,y_2;c_1,c_2,\xi)}{\partial y_1}\right|_{y_1=0,y_2=0} = -\frac{1}{\eta(\bullet)}\frac{d\eta(\bullet)}{dt}. \tag{97}$$

But on the two-dimensional plane $\{y_2=0\}\in R^3_{t,y_1,y_2}$ and along $\{y_1=0\}\cap\{y_2=0\}\in R^3_{t,y_1,y_2}$ the partial derivative $\frac{\partial f_2^2(t,y_1,y_2;c_1,c_2,\xi)}{\partial y_2}$ takes the different forms

$$\left.\frac{\partial f_2^2(t,y_1,y_2;c_1,c_2,\xi)}{\partial y_2}\right|_{y_2=0} = -\frac{1}{2}\left\{\frac{1}{\eta(\bullet)}\frac{d\eta(\bullet)}{dt}+\frac{\frac{d\gamma(\bullet)}{dt}+\frac{1}{\eta(\bullet)}\frac{d\eta(\bullet)}{dt}\gamma(\bullet)}{\gamma(\bullet)+c_2\left[y_1+\frac{c_1}{\eta(\bullet)}\right]}\right\}, \tag{98}$$

$$\left.\frac{\partial f_2^2(t,y_1,y_2;c_1,c_2,\xi)}{\partial y_2}\right|_{y_1=y_2=0} = -\frac{1}{2}\left\{\frac{1}{\eta(\bullet)}\frac{d\eta(\bullet)}{dt}+\frac{\frac{d\gamma(\bullet)}{dt}+\frac{1}{\eta(\bullet)}\frac{d\eta(\bullet)}{dt}\gamma(\bullet)}{\gamma(\bullet)+\frac{c_1c_2}{\eta(\bullet)}}\right\}. \tag{99}$$

Suppose

$$\begin{aligned}&\eta(t,\xi_1)=e^{\xi_1 t},\\&\gamma(t,\xi_2,\xi_3,\xi_4)=\xi_2+\xi_3\sin(\xi_4 t)\end{aligned} \tag{100}$$

Let us define the domain for (c_1,c_2) with the following inequalities

$$\Omega=\begin{Bmatrix}-1\le c_1\le 1\\ 0\le c_2\le 1\end{Bmatrix}. \tag{101}$$

Then we get

$$\left.\frac{\partial f_1^2(t,y_1,y_2;c_1,c_2,\xi)}{\partial y_1}\right|_{y_1=0,y_2=0} = -1, \tag{102}$$

$$\left.\frac{\partial f_2^2(t,y_1,y_2;c_1,c_2,\xi)}{\partial y_2}\right|_{y_2=0} = -\frac{1}{2}\cdot\left(1+\frac{2+\sin(t)+\cos(t)}{2+\sin(t)+c_2\left[y_1+c_1e^{-t}\right]}\right), \tag{103}$$

$$\left.\frac{\partial f_2^2(t,y_1,y_2;c_1,c_2,\xi)}{\partial y_2}\right|_{y_1=y_2=0} = -\frac{1}{2}\cdot\left(1+\frac{2+\sin(t)+\cos(t)}{2+\sin(t)+c_1c_2e^{-t}}\right) \tag{104}$$

If

$$c = \begin{vmatrix} \hat{c}_1 \\ \hat{c}_2 \end{vmatrix} \in \left\{ \begin{vmatrix} 0 \\ 0 \end{vmatrix}, \begin{vmatrix} 0.5 \\ 0.5 \end{vmatrix}, \begin{vmatrix} -0.5 \\ 0.5 \end{vmatrix} \right\} \tag{105}$$

and

$$\xi = \hat{\xi} = \begin{vmatrix} \hat{\xi}_1 \\ \hat{\xi}_2 \\ \hat{\xi}_3 \\ \hat{\xi}_4 \end{vmatrix} = \begin{vmatrix} 1 \\ 2 \\ 1 \\ 1 \end{vmatrix}, \tag{106}$$

then MATLAB simulation of the initial system I_x defined by (90) and its canonical representation I_y defined by (94) with the expressions (103) - (104) gives us the results presented by following graphs.

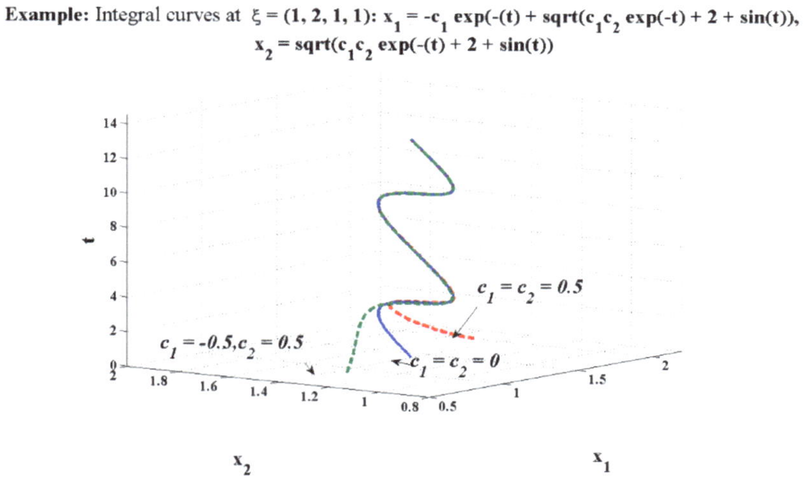

GRAPH 1

Example: Two specific sections of the corresponding two foliations defined by the first integrals

$\mathbf{g_1(t, x_1, x_2) = (x_1 - x_2)e^t = c_1}$ and $\mathbf{g_2(t, x_1, x_2) = (x_2^2 - 2 - sin(t))/(x_1 - x_2) = c_2}$ by the hyperplanes $\mathbf{(c_1 = 0.5), (c_2 = 0.5)}$

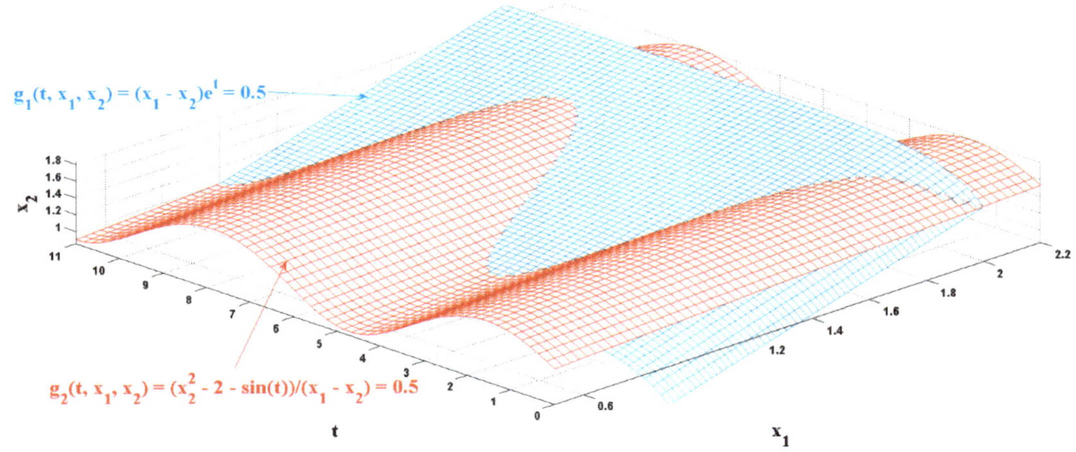

GRAPH 2

EXAMPLE: The 3D-graph of the expression of $\partial \mathbf{f_2^2(t, y_1, y_2; c, \xi)} / \partial \mathbf{y_2} \,|_{\mathbf{y_2 = 0}}$ for the analytical criterion at

$\mathbf{\xi = (\xi_1, \xi_2, \xi_3, \xi_4) = (1, 2, 1, 1), c = (c_1, c_2) \in [(0, 0)}$ *in blue,* ***(-0.5, 0.5)*** *in green,* ***(0.5, 0.5)*** *in red*].

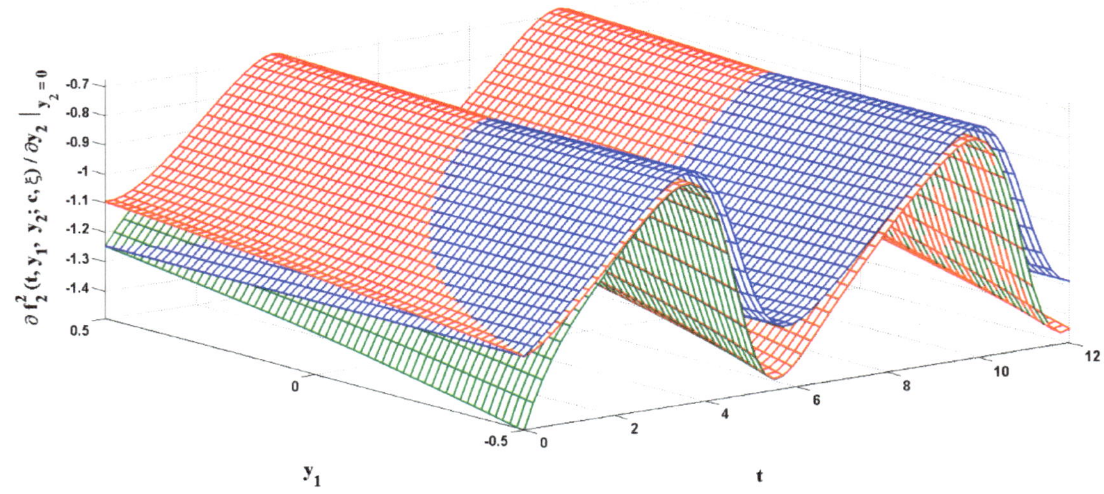

GRAPH 3

EXAMPLE: The graphs of the expression of $\partial \mathbf{f}_2^2(\mathbf{t}, \mathbf{y}_1, \mathbf{y}_2; \mathbf{c}, \xi) / \partial \mathbf{y}_2 \,|_{\mathbf{y}_2 = \mathbf{y}_1 = 0}$ for the analytical criterion at $\xi = (\xi_1, \xi_2, \xi_3, \xi_4) = (1, 2, 1, 1)$, $\mathbf{c} = (\mathbf{c}_1, \mathbf{c}_2) \in [(0, 0), (0.5, 0.5), (-0.5, 0.5)]$.

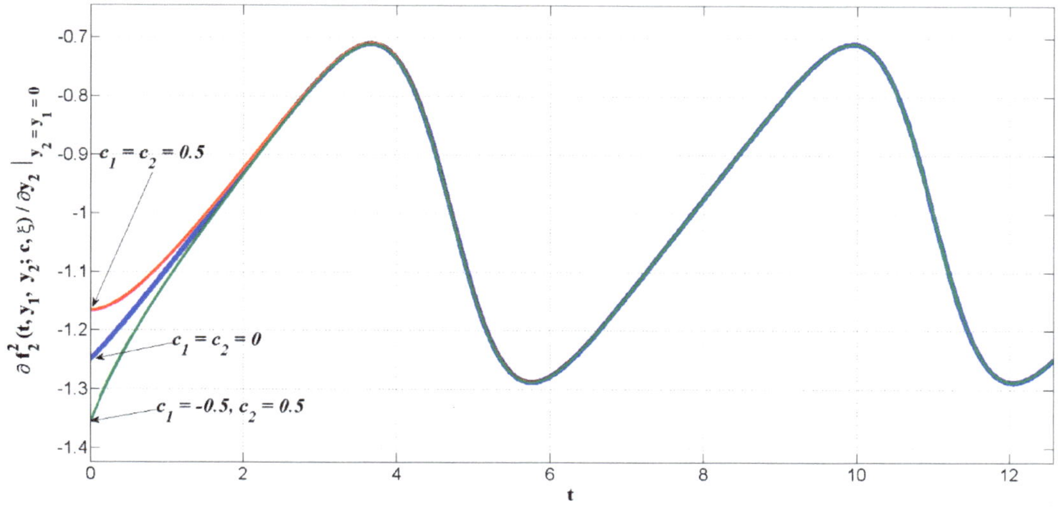

GRAPH 4

EXAMPLE: The 3D-graph of the surface $\mathbf{S}_2 \cap (\mathbf{t} = 6) = (\mathbf{s}_2 = \mathbf{f}_2^2(\mathbf{t}, \mathbf{y}_1, \mathbf{y}_2; \mathbf{c}, \xi)|_{\mathbf{t} = 6})$ being the time section of the 3-dimensional hypersurface $\mathbf{S}_2 = (\mathbf{s}_2 = \mathbf{f}_2^2(\mathbf{t}, \mathbf{y}_1, \mathbf{y}_2; \mathbf{c}, \xi))$ for the topological criterion at $\xi = (\xi_1, \xi_2, \xi_3, \xi_4) = (1, 2, 1, 1)$, $\mathbf{c} = (\mathbf{c}_1, \mathbf{c}_2) = (0, 0)$.

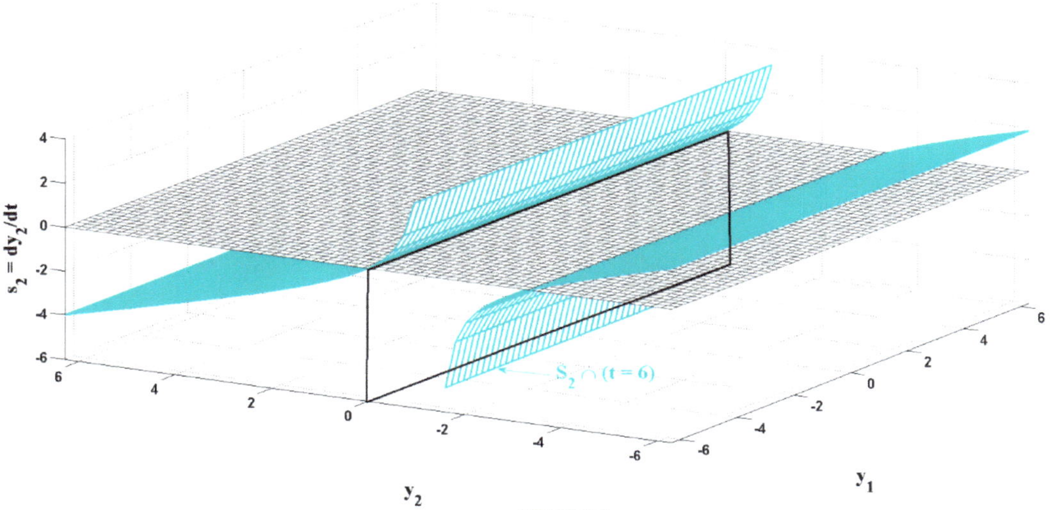

GRAPH 5

EXAMPLE: The 3D-graphs of the surfaces $\mathbf{S_2 \cap (t=0)}$ and $\mathbf{S_2 \cap (t=12)}$ being the time sections of the 3-dimensional hypersurface $\mathbf{S_2 = (s_2 = f_2^2 (t, y_1, y_2; c, \xi))}$ by the hyperplanes **(t=0)** and **(t=12)** for the topological criterion at $\xi = (\xi_1, \xi_2, \xi_3, \xi_4) = \mathbf{(1, 2, 1, 1)}$, $\mathbf{c = (c_1, c_2) = (0.5, 0.5)}$.

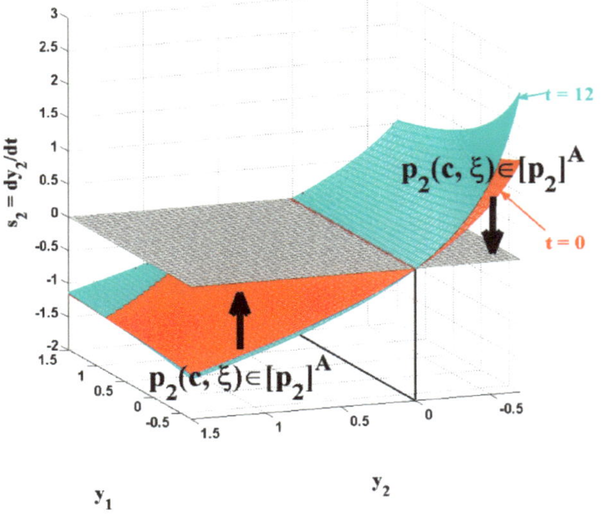

GRAPH 6

Since according to the conditions (100), (105) and (106) the canonizing diffeomorphism (92) is uniformly convergent, it is quite evident that under our conditions for the value of ξ and the values of $(c_1, c_2) \in \Omega$, there exists the canonical definite-positive auxiliary function $W(y)$ of Lyapunov functions $V(t,y)$ for the canonical system I_y in the form

$$W(y) = y_1^2 + y_2^2 \tag{107}$$

that generates the canonical positive definite Lyapunov function

$$V(t,y) = W(y)(1+e^{-t}) = (y_1^2 + y_2^2)(1+e^{-t}) \ , \tag{108}$$

ensuring the asymptotic stability of all its integral curves defined by the above-mentioned assumptions. According to Theorem 1, this function also guaranties the asymptotical stability of the corresponding integral curves or typical one-dimensional fibers of the initial system I_x.

Remark 5. *It is evident that*

$$V(t,y)\big|_{y_1=y_2=0} = (y_1^2 + y_2^2)(1+e^{-t}) = 0, \quad V(t,y)\big|_{\forall (y_1,y_2) \neq (0,0)} > 0. \tag{109}$$

The total derivative of $V(t,y)$ *is*

$$\frac{dV(t,y)}{dt} = -e^{-t}\left(y_1^2 + y_2^2\right) - 2y_1^2 + 2y_2 f_2^2\left(t, y_1, y_2; c_1, c_2, \hat{\xi}\right)\Big|_{c\in\Omega, \hat{\xi}=(1,2,1,1)}, \qquad (110)$$

where $\left.\frac{dV(t,y)}{dt}\right|_{y_1=y_2=0} = 0$.

Now we will show that the total derivative is definite-negative. Two first terms are beyond any doubts. As to the term $2y_2 f_2^2\left(t, y_1, y_2; c_1, c_2, \hat{\xi}\right)\Big|_{c\in\Omega, \hat{\xi}=(1,2,1,1)}$, *it is also negative for all* $(y_1, y_2) \neq (0,0)$ *because the covering mapping* $p_2\left(c, \hat{\xi}\right)$ *belongs to the class of* $\mathbf{A}$ *– mappings, that is* $p_2\left(c, \hat{\xi}\right) \in [p_2]^{\mathbf{A}}$. *The last statement means that for* $\xi = \hat{\xi} = (1,2,1,1)$ *and* $\forall c \in \Omega$

$$f_2^2\left(t, y_1, y_2; c_1, c_2, \hat{\xi}\right)\Big|_{y_2=0} \equiv 0 \qquad (111)$$

and

$$f_2^2\left(t, y_1, y_2; c_1, c_2, \hat{\xi}\right)\Big|_{y_2>0} < 0, \quad f_2^2\left(t, y_1, y_2; c_1, c_2, \hat{\xi}\right)\Big|_{y_2<0} > 0 \qquad (112)$$

in the quite large domain $\{1.5 > y_1 > -0.5; \quad 1.5 > y_2 > -0.5; \quad \infty > t \geq 0\}$ *(see Graph 6 as an illustration). The values of the vector of parameters* ξ *can be varied rather extensively from the chosen one of* $\hat{\xi} = (1,2,1,1)$ *but all the properties of asymptotical stability will be preserved and the qualitative results of the conducted analysis will remain unchangeable.*

5. WIDE-SENSE ROBUST, ADAPTIVE TERMINAL CONTROL FOR CIRCADIAN DYNAMICS: POINCARÉ-APPROACH-BASED BACKSTEPPING METHOD

This research has the objective to implement the differential-topological tools of the study of complex nonlinear dynamic systems for the analysis and synthesis of human-machine systems. Its history began in 2010 when Professor Guy André Boy asked me to look into a Van der Pol equation describing the jet lag and the process of adaptation after rapid transmeridian travel. After the preliminary examination I decided to add an external force to it in the attempt to build the mathematical grounds that could substantiate certain potentiality for the effective control of the process by physical, chemical and biological means. The study resulted in the paper published in 2013 [8].

However, at the beginning we would like to specify and extend the concept of robust control. Let a parametric dynamic system be governed by the following system of equations

$$\frac{dx}{dt} = f\left(t, x; \xi, \gamma, u\right) \tag{113}$$

where $t \in [t_0; +\infty[= T \subset R_t^1$ is time; $x = (x_1, ... x_n) \in R_x^n$ is a phase vector; $\xi = (\xi_1, ..., \xi_m) \in \Xi \subseteq R_\xi^m$ is a vector of system parameters and let $\xi = \xi_0 = (\xi_{1,0}, ..., \xi_{n.0}) \in \Xi$ be a nominal value of ξ; $u = \; = (u_1, ..., u_q) \in U$ is a control vector, $\dim M = q$; $\gamma = (\gamma_1, ..., \gamma_{q'}) \in \Gamma$ is a vector of undesirable external disturbances, $\dim \Gamma = q'$; $f(t, x; \xi, \gamma, u) = (f_1(t, x; \xi, \gamma, u), ..., f_n(t, x; \xi, \gamma, u)) \in C^r$, $u(t, x; a, \xi) \in C^{r-1}, r \geq 1$; $a = (a_1, ..., a_s) \in A$ is a vector of control parameters, $\dim A = s$. Let

$$x_t\left(\hat{x}_0, \hat{\xi}, \gamma, u'\right)_{|u'=u'(t,x;\hat{a},\xi_0)} = \left(t, x_1\left(t; \hat{x}_0, \hat{a}, \hat{\xi}, \gamma, u'\right)_{|u'=u'(t,x;\hat{a},\xi_0)}, ..., x_n\left(t, \hat{x}_0, \hat{a}, \hat{\xi}, \gamma, u'\right)_{|u'=u'(t,x;\hat{a},\xi_0)}\right)$$

be the integral curve of the system (113) with $\xi = \hat{\xi} \in \Xi$ affected by the external disturbances $\gamma(t)$, corresponding to some specific control law $u(t, x; \hat{a}, \xi_0)$ and starting from the initial point $\hat{x}_{t_0} = (t_0, \hat{x}_0) = (t_0, \hat{x}_{1,0}, ..., \hat{x}_{n,0}) \in T \times X_{t_0}$, where $\hat{x}_0 = (\hat{x}_{1,0}, ... \hat{x}_{n,0}) \in X_{t_0} \subseteq R_x^n$, X_{t_0} is a set of initial points of the phase vector x and $\dim X_{t_0} = n$.

Let each $x_t\left(x_0,\hat{\xi},\gamma,u'\right)\Big|_{u'=u'(t,x;\hat{a},\xi_0)}$ $\forall x_0 \in X_{t_0}$ have its ω-limit curve $x_t^{\omega}\left(x_0,\hat{\xi},\gamma,u'\right)\Big|_{u'=u'(t,x;\hat{a},\xi_0)}$:

$\lim\limits_{t\to+\infty} x_t\left(x_0,\hat{\xi},\gamma,u'\right)\Big|_{u'=u'(t,x;\hat{a},\xi_0)} = x_t^{\omega}\left(x_0,\hat{\xi},\gamma,u'\right)\Big|_{u'=u'(t,x;\hat{a},\xi_0)}$.

Denote

$M^{\omega}\left(\hat{\xi},\hat{a},\gamma\right)=\bigcup\limits_{\forall\hat{x}_0\in X_{t_0}} x_t\left(x_0,\hat{\xi},\gamma,u'\right)\Big|_{u'=u'(t,x;\hat{a},\xi_0)}$ and $M^{\omega}\left(\xi_0,\hat{a},\gamma\right)=\bigcup\limits_{\forall\hat{x}_0\in X_{t_0}} x_t\left(x_0,\xi_0,\gamma,u'\right)\Big|_{u'=u'(t,x;\hat{a},\xi_0)}$

ω-limit sets or manifolds of the system (113) with $\xi\in\left(\hat{\xi};\xi_0\right)$, where $\hat{\xi}\neq\xi_0$, $a=\hat{a}$ and $u\left(t,x;\hat{a},\xi_0\right)$. We consider the problem of terminal control formulated as follows.

TERMINAL CONTROL PROBLEM STATEMENT.

There is a given terminal manifold

$$M\left(\hat{b}\right)=\left\{\sigma_1\left(t,x;\hat{b}\right)=0,\dots,\sigma_p\left(t,x;\hat{b}\right)=0\right\}, \qquad (114)$$

which is an element of the family of terminal manifolds

$$M\left(b\right)=\left\{\sigma_1\left(t,x;b\right)=0,\dots,\sigma_p\left(t,x;b\right)=0, b\in B\right\},$$

where $1\le p\le n, \dim M=n-p+1$, $b=\left(b_1,\dots,b_{s'}\right)$ is a vector of terminal manifold parameters, $b\in B, \dim B=s'$.

The control aim is to find such a family of control laws

$$\left\{u=u'\left(t,x;a,\hat{b},\xi_0\right), a\in A\right\}$$

that

$$M^{\omega}\left(\xi_0,a,\gamma\right)\Big|_{\gamma\equiv0}=M^{\omega}\left(\xi_0,a\right)\subseteq M\left(\hat{b}\right)\forall a\in A.$$

Definition 10. The family of terminal control laws $\left\{u=u'\left(t,x;a,\hat{b},\xi_0\right), a\in A\right\}$ delivers the narrow-sense robustness to the system (113) for all $\xi\in\Xi$ if it

1) ensures the qualitative (topological) equivalency of control performance between $M^{\omega}\left(\xi,a,\gamma\right)$ and $M^{\omega}\left(\xi_0,a\right)$ for all $\xi\in\Xi, a\in A$ that is

   $M^{\omega}\left(\xi,a,\gamma\right)\approx M^{\omega}\left(\xi_0,a\right)\forall\xi\in\Xi,\forall a\in A$;

2) sustains the graph of a function of the distance $\|\cdot,\cdot\|$ between $M^{\omega}\left(\xi,a,\gamma\right)$ and $M^{\omega}\left(\xi_0,a\right)$

in the hypograph of a given non-negative majorant $\mathrm{P}\left(t, x_{p+1}, \ldots, x_n\right)$, namely

$$\left\|M^{\omega}(\xi, a, \gamma), M^{\omega}(\xi_0, a)\right\| = \rho\left(t, x_{p+1}, \ldots, x_n\right) \le \mathrm{P}\left(t, x_{p+1}, \ldots, x_n\right) \forall \xi \in \Xi, \forall a \in A,$$

where $\rho\left(t, x_{p+1}, \ldots, x_n\right) \ge 0$ is called a function of the quantitative defect of control performance, $\mathrm{P}\left(t, x_{p+1}, \ldots, x_n\right) \ge 0$, $\rho(\cdot) \in C^{r-1}, \mathrm{P}(\cdot) \in C^{r-1}, r \ge 1$.

Now let us make the constraints on u tougher shrinking the domain of admissible values by

$$u \in U' \subset U,$$

where U' is a subset of U and $\dim U' = q$.

The next step is to introduce the at least homeomorphic surjection

$$F_t : U \setminus U' \times T \to \mathrm{Fr}\, U'$$

that maps each point from the outside of U' but the inside U to the boundary of U'. This surjection can be defined by many ways and depends on the geometric form of U and the coordinate system we use. Actually this is a very interesting subject for the independent study that is not well investigated theoretically by the present time. Already as long ago as in fifties of the past century the designers synthesizing optimal control with constraints noticed the curious fact that if the actuator lacks the range of motion to execute the optimal control program obtained by calculations then at its moderate shortages the following empirical method often worked without a hitch. “Let the actuator go all way down to the positive stop following the commands of the control program and, if the next command requires it to go beyond the range of motions, let the one stay at this point until the program value comes back within the limits. After its return the actuator continues to execute the commands as before leaving the range.” It was an inexplicable intuitive sense of getting optimality in the simplest, fastest and cheapest way. The theory of nonlinear non-autonomous dynamic systems of generic configuration developed in [5] sheds the light on this empirical approach in the case of the search for stability.

Definition 11. The family of terminal control laws $\left\{u = u'\left(t, x; a, \hat{b}, \xi_0\right), a \in A\right\}$ delivers the wide-sense robustness to the system (113) if two conditions of Definition 10 are satisfied not only for all $\xi \in \Xi$ but also for $u \in U' \subset U$, where

$$\left\{ u = \begin{Bmatrix} u'\left(t, x; a, \hat{b}, \xi_0\right) & \text{if} \quad u' \in U' \\ F_t\left(u'\right) \in \mathrm{Fr}\, U' & \text{if} \quad u' \in U \setminus U' \end{Bmatrix}, a \in A \right\}.$$

Thus, the wide-sense robust control extends the concept of robustness adding the partial loss of control resources due to the sudden reduction of the control range from the nominal bounds to the uncertainty of system parameters that figures in the definition of narrow-sense robust control. Now we are ready to consider a forced van der Pol differential equation written as

$$\frac{d^2z}{dt^2} - \xi_1\left(1-z^2\right)\frac{dz}{dt} + z = u, \tag{115}$$

where z is a phase variable, t is time, ξ_1 is a system parameter, u is an external force or control. In Cauchy form the original dynamic system with control is

$$\left\{\begin{aligned} &\frac{dy}{dt} = x, \\ &\frac{dx}{dt} = \xi_1\left(1-y^2\right)x - y + u \end{aligned}\right\}, \tag{116}$$

where y and x are phase variables related to z by the equations

$$z = y, \quad \frac{dz}{dt} = \frac{dy}{dt} = x, \quad \frac{d^2z}{dt^2} = \frac{d^2y}{dt^2} = \frac{dx}{dt}.$$

The transformation $\{y = r\sin\theta, \quad x = r\cos\theta\}$ converts its look in polar variables decidedly

$$\left\{\begin{aligned} &\frac{dr}{dt} = \xi_1 \cdot r \cdot \cos^2\theta \cdot \left(1 - r^2\sin^2\theta\right) + u \cdot \cos\theta \\ &\frac{d\theta}{dt} = 1 - \xi_1 \cdot \sin\theta \cdot \cos\theta \cdot \left(1 - r^2\sin^2\theta\right) - u \cdot \frac{\sin\theta}{r} \end{aligned}\right\}, \tag{117}$$

where $r \in \mathbb{R}_{\geq 0} = [0; +\infty[$ and $\theta \in \mathbb{R}_{\geq 0}$ are phase variables; $t \in \mathbb{R}_{\geq 0}$ is time; $\xi_1 \in \Xi = \left[\xi_{1,\min}; \xi_{1,\max}\right] \subset \mathbb{R}$ is a system scalar parameter; $u \in \left[u_{\min}; u_{\max}\right] = U \subset \mathbb{R}$ is an scalar external force or control and $u = u(r, \theta; \xi_1) \in C^1\left(R^2_{r,\theta} \times \Xi \to R^1_u\right)$.

The equations (116) represent the original dynamic system with control u and phase vector (x, y). The equations (117) describe the original dynamic system with control u and phase vector (r, θ) and we will further work with the latter to design the sought control law. The equations (116) and (117) are two forms of the same original dynamic system, which are related to each other via the transformation $\{y = r\sin\theta, \quad x = r\cos\theta\}$.

The control aim is to find a terminal feedback control law $u = u'(r,\theta;\xi_1,a)$ bringing the system (117) into a terminal manifold as we will call the periodic-in-polar-variables curve of the form

$$L_{r,\theta} = \{r = g(\theta,b)\}, \tag{118}$$

where $g(\theta,b) = g(\theta + 2\pi n, b), n \in Z$; $a = (a_1,a_2) \in A \subset R^2_{a_1,a_2}$ is a vector of control parameters and $b = (b_1,b_2) \in B \subset R^2_{b_1,b_2}$ is the vector of the terminal manifold parameters. Let them belong to the space of piecewise constant functions. Designate $a_0 = (a_{1,0},a_{2,0}) \in A, b_0 = (b_{1,0},b_{2,0}) \in B, \xi_{1,0} \in \Xi$ some nominal values of $a = a_0 + \Delta a \in A, b = b_0 + \Delta b \in B, \xi_1 = \xi_{1,0} + \Delta\xi_1 \in \Xi$. The design of the sought control law will be done with $\xi_1 = \xi_{1,0}$. We also suppose that the system parameter ξ_1 can be additively and unpredictably affected by some undesirable external disturbance $\gamma_1(t)$ in the form of the bounded and at least continuous function acting only for a limited interval of time, namely

$$\left\{\gamma_1(t) = \begin{cases} \gamma_1(t) \neq 0 \forall t \in]t_1';\ t_1''[\subset T, & \text{where } \gamma_1' < \gamma_1(t) < \gamma_1'' \\ \gamma_1(t) \equiv 0 \forall t \notin]t_1';\ t_1''[\subset T & \end{cases}\right\},$$

where $\gamma_1' \in \mathbb{R},\ \ \gamma_1'' \in \mathbb{R},\ \ t_0 \le t_1' < t_1'',\ \ t_1'' < +\infty,\ \ \gamma_1(t) \in C^{r-1},\ \ r > 1.$

This means that the original system with control assumes the following look

$$\left\{\begin{aligned} \frac{dr}{dt} &= (\xi_1 + \gamma_1(t)) \cdot r \cdot \cos^2\theta \cdot (1 - r^2\sin^2\theta) + u\cos\theta \\ \frac{d\theta}{dt} &= 1 - (\xi_1 + \gamma_1(t)) \cdot \sin\theta \cdot \cos\theta \cdot (1 - r^2\sin^2\theta) - u \cdot \frac{\sin\theta}{r} \end{aligned}\right\} \tag{119}$$

There are two requirements to the control $u = \hat{u}(r,\theta,a)$. It should make the original system

- asymptotically stable in the large and wide-sense robust while the phase variable $\theta(t)$ must be a strictly monotonically increasing function of t;
- adaptive according to certain adaptation law, which will be defined later, taking into consideration the unpredictable changes of the parameters of the terminal manifold b and the system ξ_1.

Now we need to review the definition of the asymptotical stability in the large for the general case, namely for the dynamic system with control (113).

Definition 12. Let the vector function of external disturbances $\gamma(t)$ of the dynamic system with control (113) have the properties as follows:

$$\left\{ \gamma_i(t) = \begin{cases} \gamma_i(t) \neq 0 \forall t \in \left]t_i'; \;\; t_i''\right[\subset T, & \text{where} \;\; \gamma_i' < \gamma_i(t) < \gamma_i'' \\ \gamma_i(t) \equiv 0 \forall t \notin \left]t_i'; \;\; t_i''\right[\subset T & \end{cases} \right\}_{i=1}^{i=q'},$$

where $\{\gamma_i' \in \mathbb{R}, \gamma_i'' \in \mathbb{R}\}_{i=1}^{i=q'}, \{t_0 \le t_i' < t_i'', t_i'' < +\infty\}_{i=1}^{i=q'}, \gamma(t) \in C^{r-1}, r > 1$.

Its integral curve $x_t\left(\hat{x}_0, \hat{\xi}, \gamma, u'\right)\Big|_{\substack{\gamma \equiv 0 \\ u' = u'(t,x;\hat{a},\xi_0)}}$ is called asymptotically stable in the large if for any $x_0 \in X_{t_0}$ and any given ahead $\varepsilon > 0$ there exists such a $t''' > t^{\max} = \max\{t_i''\}_{i=1}^{i=q'}$ that for all $t > t'''$ the following two conditions are valid

1) $\left\| x_t\left(x_0, \hat{\xi}, \gamma, u'\right)\Big|_{u' = u'(t,x;\hat{a},\xi_0)} - x_t\left(\hat{x}_0, \hat{\xi}, \gamma, u'\right)\Big|_{\substack{\gamma \equiv 0 \\ u' = u'(t,x;\hat{a},\xi_0)}} \right\| < \varepsilon$;

2) $\lim\limits_{t \to +\infty} \left\| x_t\left(x_0, \hat{\xi}, \gamma, u'\right)\Big|_{u' = u'(t,x;\hat{a},\xi_0)} - x_t\left(\hat{x}_0, \hat{\xi}, \gamma, u'\right)\Big|_{\substack{\gamma \equiv 0 \\ u' = u'(t,x;\hat{a},\xi_0)}} \right\| = 0$.

According to the above-given definition for considering the dynamic system with control (113) asymptotically stable in the large it is not enough only to extend the neighborhood of the initial point $\hat{x}_0$ of the integral curve $x_t\left(\hat{x}_0, \hat{\xi}, \gamma, u'\right)\Big|_{\substack{\gamma \equiv 0 \\ u' = u'(t,x;\hat{a},\xi_0)}}$ to the whole domain X_{t_0} but it also requires all other integral curves $x_t\left(x_0, \hat{\xi}, \gamma, u'\right)\Big|_{u' = u'(t,x;\hat{a},\xi_0)}$ starting in X_{t_0} and affected by the external disturbances $\gamma(t)$ to converge to $x_t\left(\hat{x}_0, \hat{\xi}, \gamma, u'\right)\Big|_{\substack{\gamma \equiv 0 \\ u' = u'(t,x;\hat{a},\xi_0)}}$ at $t \to +\infty$ after the moment of time $t^{\max}$ when the disturbance $\gamma(t)$ has stopped acting on the system (113).

For control designing we will use the backstepping method based on the Poincaré's approach [9], [10]. Circa 1990 two variants of the backstepping method were developed. The first one uses Lyapunov functions to design control law [11], [12]. The second variant utilizes the tree-point Henri Poincaré's strategy for the study of manifolds and the ordinary differential equations, which right-hand sides completely define their qualitative properties. These ideas allowed to create a method of goal-oriented formation of local topological structure of co-dimension one foliations for dynamic systems with control that was laid down as the foundation of the backstepping method based on the Poincaré's approach.

STEP 1 – CHANGING PHASE VARIABLES

Let us apply the diffeomorphism

$$\left\{ r = v + g(\theta,b), \quad \theta = \theta; \quad \frac{dr}{dt} = \frac{dv}{dt} + \frac{\partial g(\theta,b)}{\partial \theta}\cdot\frac{d\theta}{dt}, \quad \frac{d\theta}{dt} = \frac{d\theta}{dt} \right\}$$

to the original system (117) with $\xi_1 = \xi_{1,0}$.

We receive a new transformed system

$$\left\{ \begin{aligned} &\frac{dv}{dt} = P(v,\theta,u,\xi_{1,0},b) \\ &\frac{d\theta}{dt} = 1 - \xi_{1,0}\cdot\sin\theta\cdot\cos\theta\cdot\left(1-(v+g(\theta))^2\sin^2\theta\right) - u\cdot\frac{\sin\theta}{(v+g(\theta))} \end{aligned} \right\},$$

where $P(v,\theta,u,\xi_{1,0},b) = u\cdot\left(\cos\theta + \frac{\sin\theta}{(v+g(\theta,b))}\cdot\frac{\partial g(\theta,b)}{\partial\theta}\right) + \xi_{1,0}\cdot\cos^2\theta\cdot(v+g(\theta,b))\times$

$$\times\left[1-(v+g(\theta,b))^2\cdot\sin^2\theta\right] + \xi_{1,0}\cdot\sin\theta\cdot\cos\theta\cdot\frac{\partial g(\theta,b)}{\partial\theta}\cdot\left[1-(v+g(\theta,b))^2\sin^2\theta\right] - \frac{\partial g(\theta,b)}{\partial\theta}.$$

STEP 2 - FORMING THE RIGHT-HAND SIDE OF THE DIFFERENTIAL EQUATION

We set the equation

$$P(v,\theta,u,\xi_{1,0},b) = \chi(v,\theta,a).$$

Solving it for u, we obtain the sought control law in the form

$$u = \bar{u}(v,\theta,a,b,\xi_{1,0}) = \frac{\chi(v,\theta,a) - \bar{\Gamma}_1\cdot\xi_{1,0}\cdot\cos\theta\cdot\bar{\Gamma}_2 + \dfrac{\partial g(\theta,b)}{\partial\theta}}{\cos\theta + \dfrac{\partial g(\theta,b)}{\partial\theta}\cdot\dfrac{\sin\theta}{(v+g(\theta,b))}}, \tag{120}$$

where $\bar{\Gamma}_1 = 1-(v+g(\theta,b))^2\sin^2\theta, \quad \bar{\Gamma}_2 = \cos\theta\cdot(v+g(\theta,b)) + \sin\theta\cdot\frac{\partial g(\theta,b)}{\partial\theta}$.

Thus the canonical form of the representation of the initial system is

$$\left\{ \begin{aligned} &\frac{dv}{dt} = \chi(v,\theta,a) \\ &\frac{d\theta}{dt} = 1 - \xi_{1,0}\cdot\sin\theta\cdot\cos\theta\cdot\left(1-(v+g(\theta))^2\sin^2\theta\right) - \bar{u}(v,\theta,a,b,\xi_{1,0})\cdot\frac{\sin\theta}{(v+g(\theta))} \end{aligned} \right\}. \tag{121}$$

STEP 3 - RETURNING TO OLD PHASE VARIABLES AND OBTAINING THE TRIAD OF "INITIAL FORM<=>CANONIZING DIFFEOMORPHISM<=>CANONICAL FORM"

Using the inverse transformation $\{v = r - g(\theta, b),\quad \theta = \theta\}$ we return to the initial system (117)

$$u = u'(r,\theta,a,b,\xi_{1,0}) = \frac{\left\{\chi(r-g(\theta,b),\theta,a) - \Gamma_1\cdot\xi_{1,0}\cdot\cos\theta\cdot\Gamma_2 + \frac{\partial g(\theta,b)}{\partial\theta}\right\}}{\left\{\cos\theta + \frac{\partial g(\theta,b)}{\partial\theta}\frac{\sin\theta}{r}\right\}},$$

where $\Gamma_1 = 1 - r^2\sin^2\theta,\quad \Gamma_2 = r\cdot\cos\theta + \sin\theta\frac{\partial g(\theta)}{\partial\theta}$.

We have the triad of the initial system, its canonical form and the canonizing diffeomorphism.

$$I_{r,\theta} = \boxed{\begin{array}{l} \frac{dr}{dt} = (\xi_1 + \gamma_1(t))\cdot r\cdot\cos^2\theta\cdot(1 - r^2\sin^2\theta) + u\cdot\cos\theta \\ \frac{d\theta}{dt} = 1 - (\xi_1+\gamma_1(t))\cdot\sin\theta\cdot\cos\theta(1-r^2\sin^2\theta) - u\cdot\frac{\sin\theta}{r} \\ u = u'(r,\theta;a,b,\xi_{1,0}) = \frac{\left\{\chi(r-g(\theta,b),\theta,a) - \Gamma_1\cdot\xi_{1,0}\cdot\cos\theta\cdot\Gamma_2 + \frac{\partial g(\theta,b)}{\partial\theta}\right\}}{\left\{\cos\theta + \frac{\partial g(\theta,b)}{\partial\theta}\frac{\sin\theta}{r}\right\}} \\ \Gamma_1 = 1 - r^2\sin^2\theta,\quad \Gamma_2 = r\cdot\cos\theta + \sin\theta\frac{\partial g(\theta)}{\partial\theta} \end{array}} \tag{122}$$

$$\Updownarrow$$

$$\overline{\varphi}(\overline{\varphi}^{-1}) = \boxed{\{v = r - g(\theta,b), \theta = \theta\} : L_{r,\theta} = \{r = g(\theta,b)\} \xleftrightarrow{\overline{\varphi}(\overline{\varphi}^{-1})} L_{v,\theta} = \{v = 0\}} \tag{123}$$

$$\Updownarrow$$

$$I_{v,\theta} = \boxed{\begin{array}{l} \frac{dv}{dt} = \chi(v,\theta,a) \\ \frac{d\theta}{dt} = 1 - (\xi_1+\gamma_1(t))\cdot\sin\theta\cdot\cos\theta\cdot\left(1 - (v+g(\theta))^2\sin^2\theta\right) - \overline{u}\cdot\frac{\sin\theta}{(v+g(\theta))} \\ u = \overline{u}(v,\theta;a,b,\xi_{1,0}) = \frac{\chi(v,\theta,a) - \overline{\Gamma}_1\cdot\xi_{1,0}\cdot\cos\theta\cdot\overline{\Gamma}_2 + \frac{\partial g(\theta,b)}{\partial\theta}}{\cos\theta + \frac{\partial g(\theta,b)}{\partial\theta}\frac{\sin\theta}{(v+g(\theta,b))}} \\ \overline{\Gamma}_1 = 1 - (v+g(\theta,b))^2\sin^2\theta,\quad \overline{\Gamma}_2 = (v+g(\theta,b))\cos\theta + \sin\theta\frac{\partial g(\theta,b)}{\partial\theta} \end{array}} \tag{124}$$

It is necessary to keep in mind two things. First, $I_{r,\theta}$ and $I_{v,\theta}$ are five-parametric differential inclusions with the unified vector of parameters $(\xi_1, a_1, a_2, b_1, b_2)$. Second, the canonizing diffeomorphism $\bar{\varphi}(\bar{\varphi}^{-1})$ establishes the bijective smooth map between the terminal manifold $L_{r,\theta} = \{r = g(\theta,b)\}$ of $I_{r,\theta}$ and the terminal manifold $L_{v,\theta} = \{v = 0\}$ of $I_{v,\theta}$.

STEP 4 - DEFINING THE FUNCTION $\boldsymbol{\chi(v,\theta,a)}$ THAT VALIDATES CANONICAL LYAPUNOV FUNCTIONS ENSURING PARTIAL STABILITY FOR THE SYSTEM

According to [5] to ensure robustness and stability in the large the function $\chi(v,\theta,a)$ should be associated with the covering mapping π_v belonging to the class of the covering maps $[\pi_v]^A$ with $rank\pi_v = 1$, which create structurally-stable attractors in phase spaces. The parametric function

$$\chi = \chi(v,\theta,a) = -a_2 \cdot \arctan(a_1 v) \tag{125}$$

satisfies the condition and even moreover it allows to take account of the constraints imposed on the control u.

With such a function $\chi(v,\theta,a)$ the canonizing diffeomorphism $\bar{\varphi}(\bar{\varphi}^{-1}) = \{v = r - g(\theta,b), \theta = \theta\}$ of the triad

$$I_{r,\theta} \xleftrightarrow{\bar{\varphi}(\bar{\varphi}^{-1})} I_{v,\theta} \tag{126}$$

remains unchanged but its differential inclusions assumes the following forms respectively

$$I_{r,\theta} = \boxed{\begin{array}{l} \dfrac{dr}{dt} = (\xi_1 + \gamma_1(t)) \cdot r \cdot \cos^2\theta \cdot (1 - r^2 \sin^2\theta) + u \cdot \cos\theta \\ \dfrac{d\theta}{dt} = 1 - (\xi_1 + \gamma_1(t)) \cdot \sin\theta \cdot \cos\theta \cdot (1 - r^2 \sin^2\theta) - u \cdot \dfrac{\sin\theta}{r} \\ u = u'(r,\theta; a, b, \xi_{1,0}) = \dfrac{\left\{-a_2 \cdot \arctan\left[a_1 (r - g(\theta,b))\right] - \Gamma_1 \cdot \xi_{1,0} \cdot \cos\theta \cdot \Gamma_2 + \dfrac{\partial g(\theta,b)}{\partial \theta}\right\}}{\left\{\cos\theta + \dfrac{\partial g(\theta,b)}{\partial \theta} \cdot \dfrac{\sin\theta}{r}\right\}} \\ \\ \Gamma_1 = 1 - r^2 \sin^2\theta, \quad \Gamma_2 = r \cdot \cos\theta + \sin\theta \cdot \dfrac{\partial g(\theta,b)}{\partial \theta} \end{array}} \tag{127}$$

and

$$
I_{v,\theta} = \left[\begin{array}{l}
\dfrac{dv}{dt} = -a_2 \cdot \arctan\left(a_1 v\right) \\
\dfrac{d\theta}{dt} = 1 - \left(\xi_1 + \gamma_1(t)\right) \cdot \sin\theta \cdot \cos\theta \cdot \left(1 - \left(v + g(\theta)\right)^2 \sin^2\theta\right) - u \cdot \dfrac{\sin\theta}{\left(v + g(\theta)\right)} \\
u = \bar{u}\left(v, \theta; a, b, \xi_{1,0}\right) = \dfrac{-a_2 \cdot \arctan\left(a_1 v\right) - \bar{\Gamma}_1 \cdot \xi_{1,0} \cdot \cos\theta \cdot \bar{\Gamma}_2 + \dfrac{\partial g(\theta, b)}{\partial\theta}}{\cos\theta + \dfrac{\partial g(\theta, b)}{\partial\theta} \dfrac{\sin\theta}{\left(v + g(\theta, b)\right)}} \\
\bar{\Gamma}_1 = 1 - \left(v + g(\theta, b)\right)^2 \sin^2\theta, \quad \bar{\Gamma}_2 = \cos\theta \cdot \left(v + g(\theta, b)\right) + \sin\theta \dfrac{\partial g(\theta, b)}{\partial\theta}
\end{array} \right]. \tag{128}
$$

According to our problem statement the phase variable $\theta(t)$ must belong to the class of strictly monotonically increasing functions but at the same time the phase variable $r(\theta(t))$ must be asymptotically stable and converge to the curve $M = \{r = g(\theta(t), b)\}$. In other words we need to establish the asymptotical stability with respect to r only "conniving" at the instability of θ. This means that our case falls under the class of the problems of partial stability or stability with respect to part of variables [13], [14].

Now we will investigate the canonical form $I_{v,\theta}$ of the representation of our differential inclusion for stability only with respect to the component of v. First, we choose a canonical auxiliary Lyapunov function. Let it be

$$
W(v) = v^2. \tag{129}
$$

The v-positive-definite Lyapunov function can be written, for example, as follows

$$
V(t, v, \theta) = v^2\left(\lambda + e^{-\theta} + e^{-t}\right) \geq W(v), \tag{130}
$$

where $\lambda \in \mathbb{R}^+$ and $\lambda \geq 1$, $\{\theta, t\} \in R_2^{\geq 0}$, $\left(\lambda + e^{-\theta} + e^{-t}\right) > 1$.

We have $\dfrac{dV}{dt} = 2v\dfrac{dv}{dt}\left\{\lambda + e^{-\theta} + e^{-t}\right\} - v^2\left\{e^{-\theta}\dfrac{d\theta(t)}{dt} + e^{-t}\right\}$, where $\dfrac{dv}{dt} = -a_2 \cdot \arctan\left(a_1 v\right)$.

It is obvious that $V(t, v, \theta)$ satisfies the conditions of Theorem 0.4.2 [14] naturally because always such a real number $\mu > 0$ with the corresponding domain $H = H^- \cup H^0 \cup H^+$ can be found that for $\forall\, v \in H$ we have

- $a(\|v\|)=\|v\|^2 \le V(t,v,\theta) \le b(\|v\|)=\|v\|^2(\lambda+2);$
- $\dfrac{dV}{dt}=-2a_2v\cdot\arctan(a_1v)\{\lambda+e^{-\theta}+e^{-t}\}-v^2\left\{e^{-\theta}\dfrac{d\theta(t)}{dt}+e^{-t}\right\}\le$

$$\le -2a_2v\cdot\arctan(a_1v)\{\lambda+e^{-\theta}+e^{-t}\}\le 2v\cdot\overbrace{\{-a_2\cdot\arctan(a_1v)\}}^{\boxed{\chi(v,\theta,a)\left\{\begin{matrix}<-\mu v\forall v\in H^+\\>-\mu v\forall v\in H^-\end{matrix}\right\}}} \le -2\mu\cdot\overbrace{\{v^2\}}^{\boxed{\|v\|^2}}=-c(\|v\|),$$

where $e^{-\theta}>0\forall\theta\in\mathbb{R}$, $e^{-t}>0\forall t\in T$, $\left\{v\dfrac{dv}{dt}=-2a_2v\cdot\arctan(a_1v)<0,\dfrac{d\theta(t)}{dt}>0\right\}\forall t\in T$, the functions $\{a(\|v\|),\ b(\|v\|),\ c(\|v\|)\}\in C^0\left(R^1_{\|v\|}\to\mathbb{R}_{\ge 0}\right)$ belong to the class of strictly and monotonically increasing functions of $\|v\|$ and $a(0)=b(0)=c(0)=0$, $\|\vartheta\|=\sqrt{\sum_{i=1}^{n}\vartheta_i^2}$ for a vector $\vartheta=\begin{vmatrix}\vartheta_1\\ \dots\\ \vartheta_n\end{vmatrix}\Rightarrow\|v\|=\sqrt{v^2}=|v|$ (see Graph 7 shown below illustrating $\chi(v,\theta,a)\left\{\begin{matrix}<-\mu v\forall v\in H^+\\>-\mu v\forall v\in H^-\end{matrix}\right\}$).

The graphs of the controlling function $\chi=\chi(v,\theta,a)=-a_2\cdot\arctan(a_1\cdot v)$ with $a_2=3$, $a_1=9$ and its minorant/majorant function $\chi=-\mu\cdot v$ in the domain $H=H^-\cup H^0\cup H^+$, where $\mu=1$

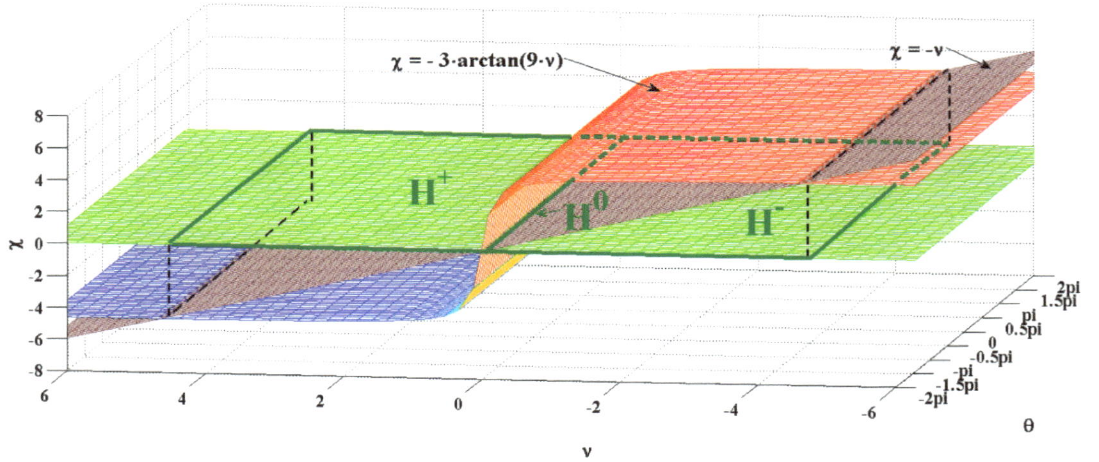

GRAPH 7

The straight line $\{\chi = -\mu \cdot \nu\}$ is a minorant for $\forall \nu \in H^-$ and majorant for $\forall \nu \in H^+$ of the function of $\chi = -a_2 \cdot \arctan(a_1 \nu)$. Thus the terminal manifold $L_{\nu,\theta} = \{\nu = 0\}$ of $I_{\nu,\theta}$ is asymptotically stable.

Now turn our attention to the canonizing diffeomorphism $\bar{\varphi}^{-1}(\bar{\varphi}) = \{r = \nu + g(\theta, b), \quad \theta = \theta\}$.

Because of the linear dependence between r and ν the parametric function $r = \nu + g(\theta, b)$ is uniformly convergent in ν to $r = g(\theta, b)$, namely

$$\{r = \nu + g(\theta, b)\} \overbrace{- - - \to}^{\nu \to 0} \{r = g(\theta, b)\}. \tag{131}$$

This fact implies that the second condition of Theorem 1 is also met and this leads us to the conclusion that the terminal manifold $L_{r,\theta} = \{r = g(\theta, b)\}$ of the differential inclusion $I_{r,\theta}$ is asymptotically stable too. Thus we have proved that the control law

$$\left\{\begin{aligned} & u = u'(r, \theta; a, b, \xi_{1,0}) = \frac{\left\{-a_2 \cdot \arctan\left[a_1 \left(r - g(\theta, b)\right)\right] - \Gamma_1 \cdot \xi_{1,0} \cdot \cos\theta \cdot \Gamma_2 + \dfrac{\partial g(\theta, b)}{\partial \theta}\right\}}{\left\{\cos\theta + \dfrac{\partial g(\theta, b)}{\partial \theta} \dfrac{\sin\theta}{r}\right\}}, \\ & \text{where} \quad \Gamma_1 = 1 - r^2 \sin^2\theta, \quad \Gamma_2 = r \cdot \cos\theta + \sin\theta \frac{\partial g(\theta, b)}{\partial \theta} \end{aligned}\right\} \tag{132}$$

is the solution to our problem of terminal control of the original system (117).

STEP 5 – CHOOSING THE SYSTEM PARAMETER AND THE TERMINAL MANIFOLD

Let the nominal system parameter be $\xi_1 = \xi_{1,0} = 0.1$ and the terminal manifold be as

$$L_{r,\theta} = \{r = g(\theta, b) = b_2 + b_1 \sin(\theta)\}, \tag{133}$$

here (b_2, b_1) are parameters and $b_2 > b_1 > 0, (b_2, b_1) \in R^2$. The following two graphs illustrate the same terminal manifold with $b_2 = b_{2,0} = 4, b_1 = b_{1,0} = 1.5$ in orthogonal Cartesian coordinate system with two different sets of coordinates, namely (x, y) and (r, θ).

We need once again to emphasize the very important point to avoid any ambiguity. The original dynamic system with control written either in the form (116) or in the form (117) is always considered in orthogonal Cartesian coordinate system as well its terminal manifold $L_{r,\theta}$. Both these two forms are related to each other through the diffeomorphism $\{y = r\sin\theta, \quad x = r\cos\theta\}$

that is in fact the transformation between orthogonal Cartesian and polar coordinates systems. The difference between two forms of the original dynamic system with control u lies only in the phase vectors (x, y) and (r, θ).

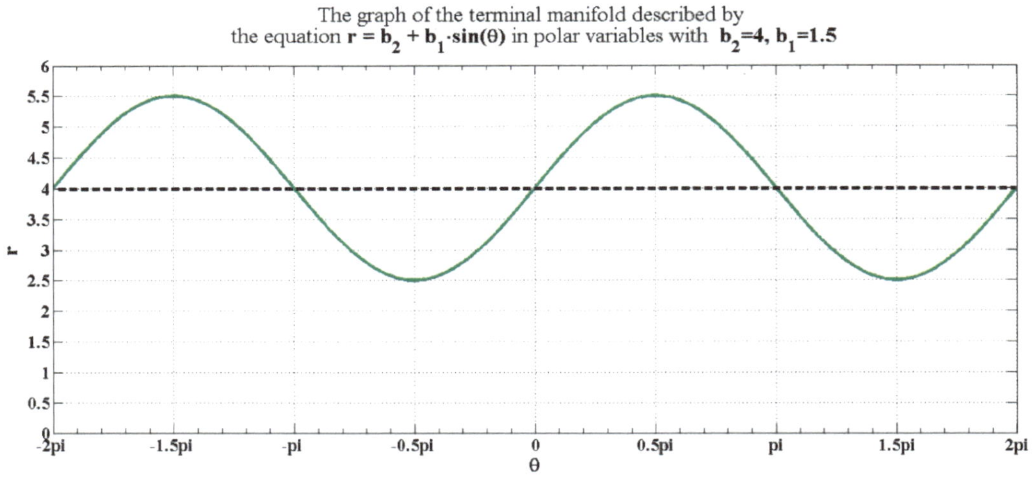
The graph of the terminal manifold described by
the equation r = b2 + b1·sin(θ) in polar variables with b2=4, b1=1.5
r
θ
-2pi
-1.5pi
-pi
-0.5pi
0
0.5pi
pi
1.5pi
2pi

GRAPH 8

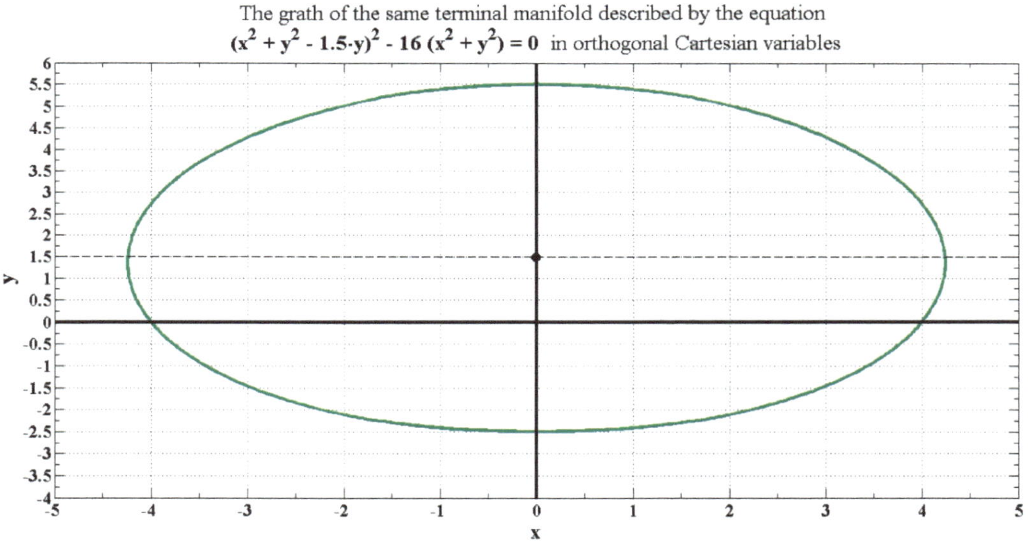
The grath of the same terminal manifold described by the equation
(x2 + y2 - 1.5·y)2 - 16 (x2 + y2) = 0 in orthogonal Cartesian variables
y
x

GRAPH 9

The surface representing the function (125) with the specific parameters $a_1 = a_{1,0} = 9, a_2 = a_{2,0} = 3$ has been given above on Graph 7.

STEP 6 - COMPUTER SIMULATION IN MATLAB

Let, first, the system parameter be nominal $\xi_1 = \xi_{1,0} = 0.1$, the terminal manifold be $M = \{r = g(\theta, b) = b_2 + b_1 \sin(\theta)\}$ with the parameters $b_2 = 4, b_1 = 1.5$. The control parameters are $a_1 = 9, \quad a_2 = 3$. Thus, we will execute the computer simulation of the original system (117)

$$\left\{ \begin{aligned} \frac{dr}{dt} &= (\xi_1 + \gamma_1(t)) \cdot r \cdot \cos^2\theta \cdot (1 - r^2 \sin^2\theta) + u \cdot \cos\theta \\ \frac{d\theta}{dt} &= 1 - (\xi_1 + \gamma_1(t)) \cdot \sin\theta \cdot \cos\theta \cdot (1 - r^2 \sin^2\theta) - u \cdot \frac{\sin\theta}{r} \end{aligned} \right\}$$

with $\xi_1 = \{0.08; \quad \xi_{1,0} = 0.1; \quad 0.12\}$ governed by the control law

$$\left\{ u = \frac{\{-3 \cdot \arctan[9 \cdot (r - 4 - 1.5 \cdot \sin\theta)] - \xi_{1,0} \cdot \cos^2\theta \cdot (1 - r^2 \sin^2\theta) \cdot (r + 1.5 \cdot \sin\theta) + 1.5 \cdot \cos\theta\}}{\cos\theta \left\{1 + 1.5 \cdot \dfrac{\sin\theta}{r}\right\}} \right\},$$

which should bring the moving phase point $(r(t), \theta(t))$ into $M = \{r = 4 + 1.5 \cdot \sin(\theta)\}$.

Thus, the values of the system parameter ξ_1 will be varied within the set $\{0.08; \quad 0.1; \quad 0.12\}$.

The computer simulation is illustrated with the graphs given below in three sections. The first one demonstrates the asymptotical stability in the large of the original dynamic system with control governed by the designed control law. The second section shows its asymptotical stability in the large together with the narrow-sense robustness when the system parameter floats within the given limits. The last third section numerically validates the asymptotical stability in the large and the wide-sense robustness of the original system in the case when the uncertainty and volatility of the control parameter is aggravated with the partial loss of control resource. If it is not qualified in particular cases, the red color of any 2D curve means that the original dynamic system with control has started from the initial point $(r(0) = 4, \theta(0) = 0) \Leftrightarrow (x(0) = 4, y(0) = 0)$, the green color corresponds to the initial point $(r(0) = 6, \theta(0) = 0) \Leftrightarrow (x(0) = 6, y(0) = 0)$, and the blue color represents the initial point $(r(0) = 2, \theta(0) = 0) \Leftrightarrow (x(0) = 2, y(0) = 0)$.

The majorant $\mathrm{P}(\theta) = const$ of the function of the quantitative defect of control performance $\rho(\theta)$ is different for the narrow-sense and wide-sense cases of robustness. For the former it is 0.26, for the latter it equals 0.7. The similar skewness has been also accepted for the partial loss of control resources. If the system parameter ξ_1 assumes the lower value 0.05 then the range of the admissible values of control u drops from its nominal interval $[-10;\ \ 10]$ to $[-2;\ \ 2]$. If it takes on the upper value 0.2 then the range reduces to $[-7;\ \ 7]$. This incommensurability is explained by the asymmetric response of the original system to all the kinds of changes mostly because of the asymmetric position of the terminal manifold with reference to the coordinate axis x of the orthogonal Cartesian system. The exposure of the system to external disturbances acting for a limited period of time has been simulated in two ways. In order to test the asymptotical stability in the large we activate the disturbance $\gamma_1(t)$ at the very first moment of the motion of the system. In order to assess the narrow-sense and wide-sense types of robustness we launch it closer to the middle of the motion. The asymptotical stability in the large requires the system to return to the undisturbed form of the motion after the disturbing factor has stopped affecting it.

Substep 6.1 – Asymptotical Stability in the Large

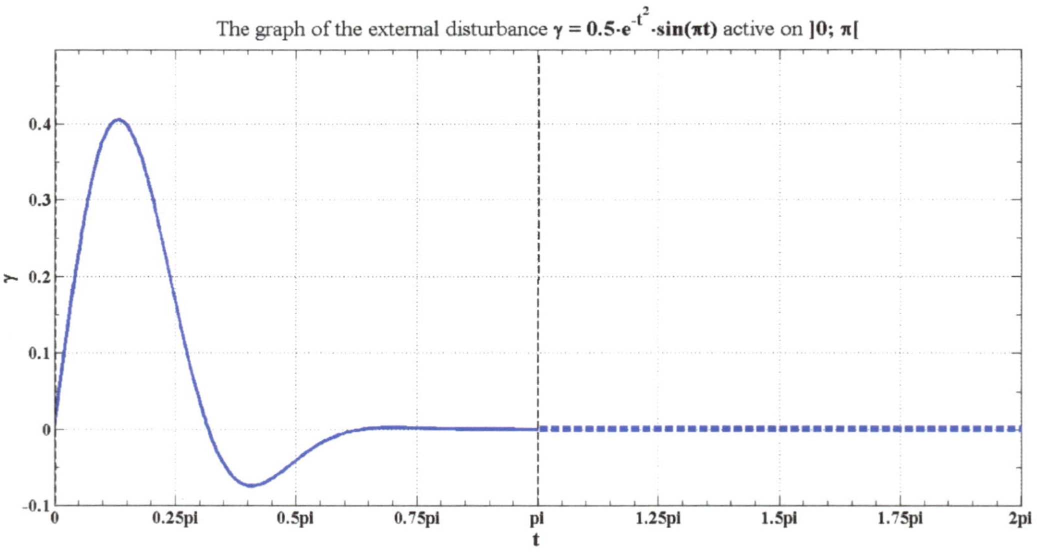

GRAPH 10

THE ASYMPTOTICAL STABILITY IN THE LARGE of the original system with phase vector **(r, θ)**: the component **r(t)** of integral curves is in blue, green and red with $\xi_1 = \xi_{1,0} = \mathbf{0.1}$ without as continuous lines and with as dashed lines the external disturbance $\boldsymbol{\gamma(t) = 0.5 \cdot \exp(-t^2)\ \sin(\pi \cdot t) \forall t \in]0;\ \pi[\ \&\ 0 \forall t \geq \pi}$

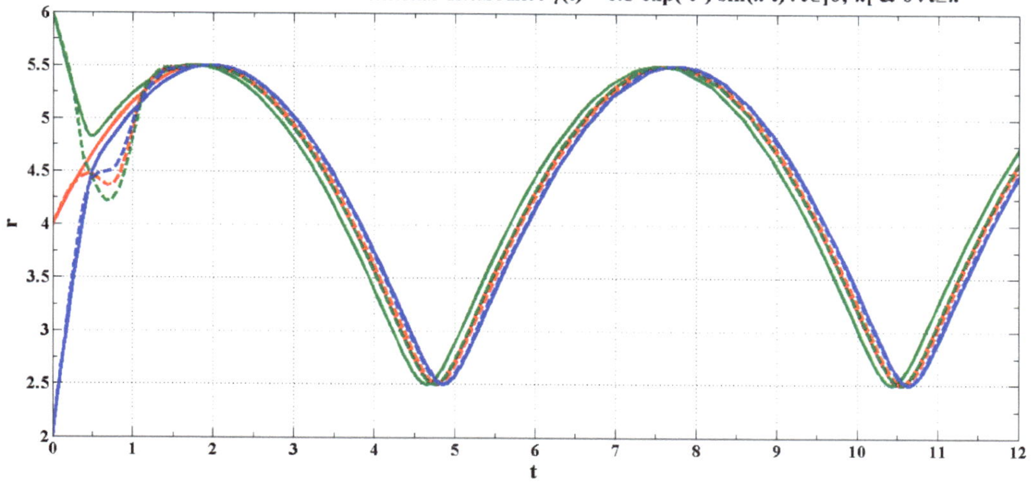

GRAPH 11

THE ASYMPTOTICAL STABILITY IN THE LARGE of the original system with phase vector **(r, θ)**: the terminal manifold $\mathbf{\{r = 4 + 1.5 \cdot \sin(\theta)\}}$ is in red and the phase trajectories are in blue, green and red with $\xi_1 = \xi_{1,0} = \mathbf{0.1}$, without as continuous lines and with as dashed lines the external disturbance

$\boldsymbol{\gamma(t) = 0.5 \cdot \exp(-t^2)\ \sin(\pi \cdot t) \forall t \in]0;\ \pi[\ \&\ 0 \forall t \geq \pi}$

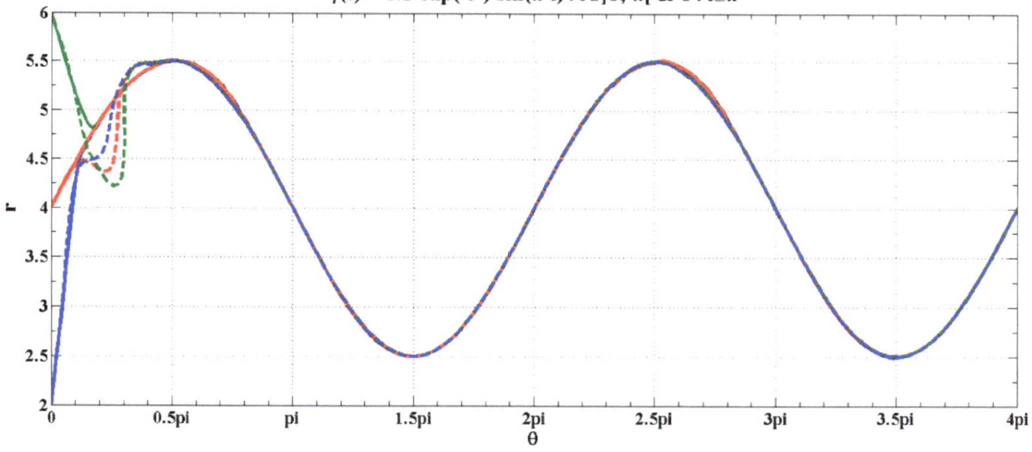

GRAPH 12

THE ASYMPTOTICAL STABILITY IN THE LARGE of the original system with phase vector **(r, θ)**: the component **θ(t)** of integral curves is in blue, green and red with $\xi_1 = \xi_{1,0} = \mathbf{0.1}$, without as continuous lines and with as dashed lines the external disturbance $\boldsymbol{\gamma(t) = 0.5\cdot\exp(-t^2)\,\sin(\pi\cdot t)\forall t\in]0;\,\pi[\ \&\ 0\forall t\geq\pi}$ on a large scale

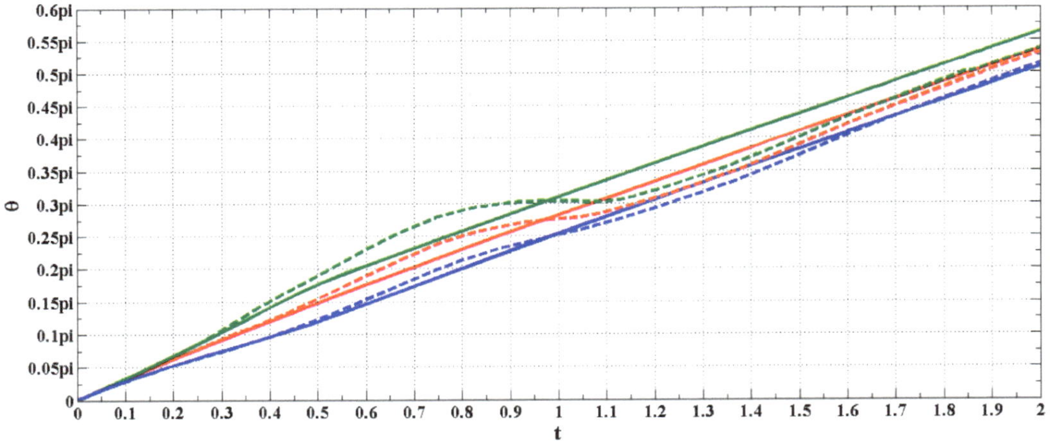

GRAPH 13

THE ASYMPTOTICAL STABILITY IN THE LARGE of the original system with phase vector **(r, θ)**: the component **θ(t)** of integral curves is in blue, green and red with $\xi_1 = \xi_{1,0} = \mathbf{0.1}$, without as continuous lines and with as dashed lines the external disturbance $\boldsymbol{\gamma(t) = 0.5\cdot\exp(-t^2)\,\sin(\pi\cdot t)\forall t\in]0;\,\pi[\ \&\ 0\forall t\geq\pi}$ on a small scale

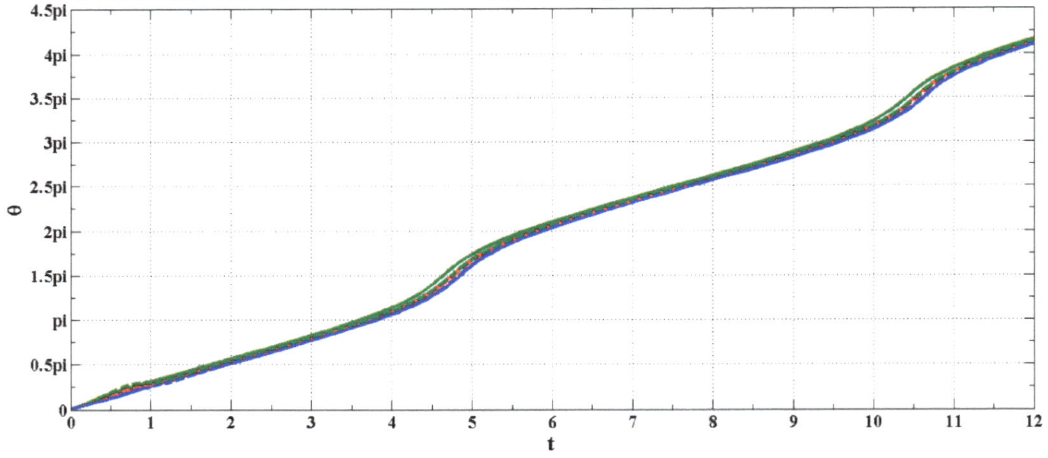

GRAPH 14

THE ASYMPTOTICAL STABILITY IN THE LARGE of the original system with phase vector **(r, θ)**:
the control **u(θ)** is in blue, green and red with $\xi_1 = \xi_{1,0} = \mathbf{0.1}$ without as continuous lines and with
as dashed lines the external disturbance $\boldsymbol{\gamma(t) = 0.5 \cdot \exp(-t^2)\,\sin(\pi \cdot t) \forall t \in]0;\, \pi[\ \&\ 0 \forall t \geq \pi}$

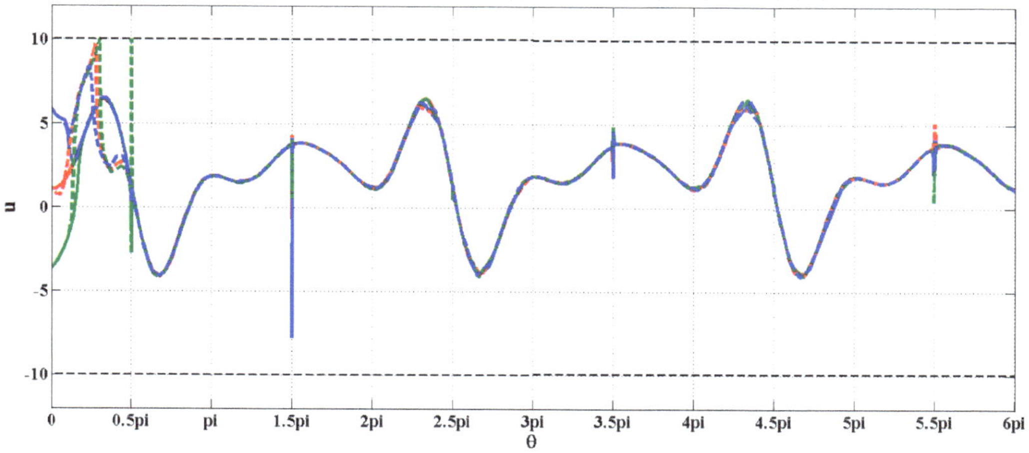

GRAPH 15

THE ASYMPTOTICAL STABILITY IN THE LARGE of the original system with phase vector **(x, y)**
is presented by the integral curves **(x(t), y(t))** (**x** in red, brown, purple and **y** in green, black, blue) with $\xi_1 = \xi_{1,0} = \mathbf{0.1}$
without as continuous lines and with as dashed lines the external disturbance $\boldsymbol{\gamma(t) = 0.5 \cdot \exp(-t^2)\,\sin(\pi \cdot t) \forall t \in]0;\, \pi[\ \&\ 0 \forall t \geq \pi}$
on a large scale

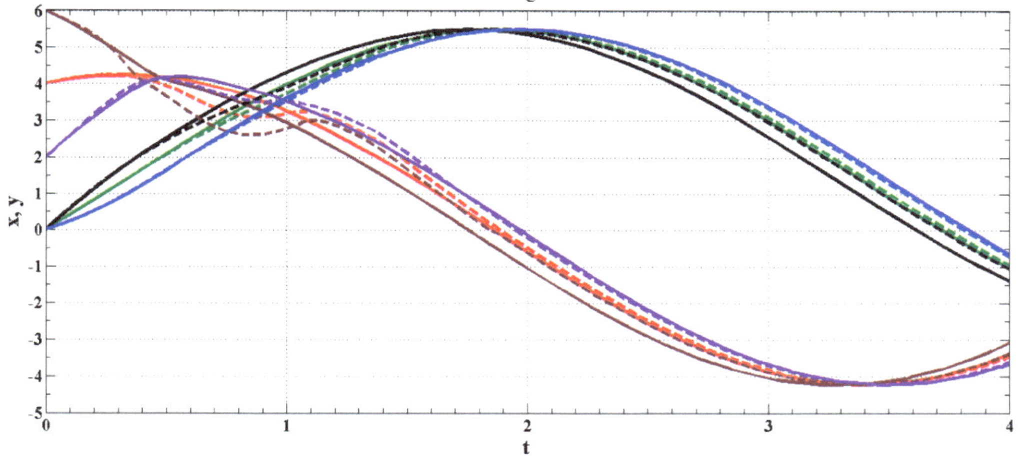

GRAPH 16

THE ASYMPTOTICAL STABILITY IN THE LARGE of the original system with phase vector **(x, y)** is presented by the integral curves **(x(t), y(t))** (**x** in red, brown, purple and **y** in green, black, blue) with $\xi_1 = \xi_{1,0} = 0.1$ without as continuous lines and with as dashed lines the external disturbance $\gamma(t) = 0.5 \cdot \exp(-t^2)\ \sin(\pi \cdot t) \forall t \in]0;\ \pi[\ \&\ 0 \forall t \geq \pi$ on a small scale

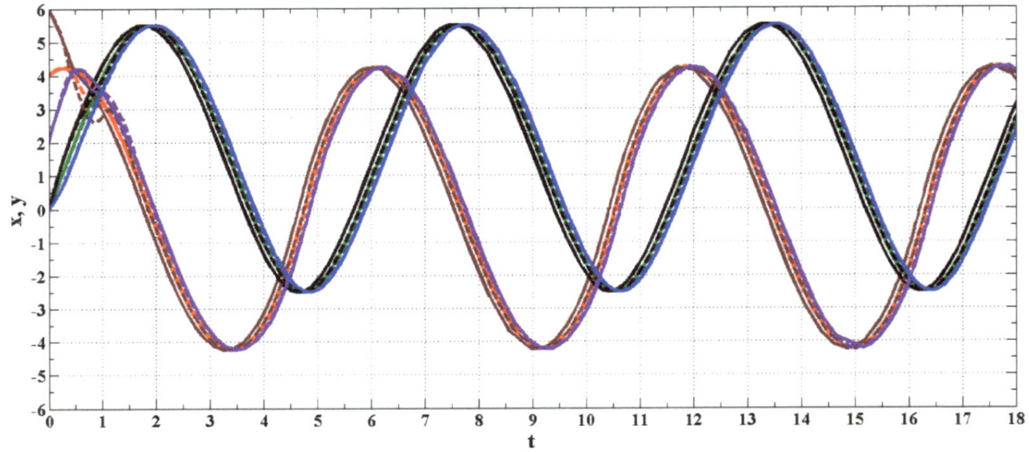

GRAPH 17

THE ASYMPTOTICAL STABILITY IN THE LARGE of the original system with phase vector **(x, y)**: the terminal manifold $(x^2 + y^2 - 1.5\ y)^2 - 16\ (x^2 + y^2) = 0$ is in yellow dots, the phase orbits are in blue, green and red with $\xi_1 = \xi_{1,0} = 0.1$ and the external disturbance

$\gamma(t) = 0.5 \cdot \exp(-t^2)\ \sin(\pi \cdot t) \forall t \in]0;\ \pi[\ \&\ 0 \forall t \geq \pi$

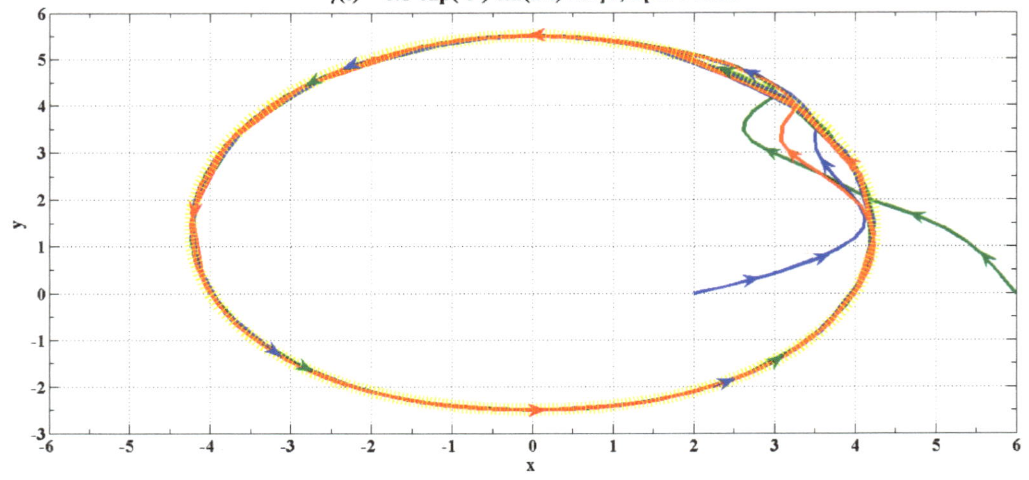

GRAPH 18

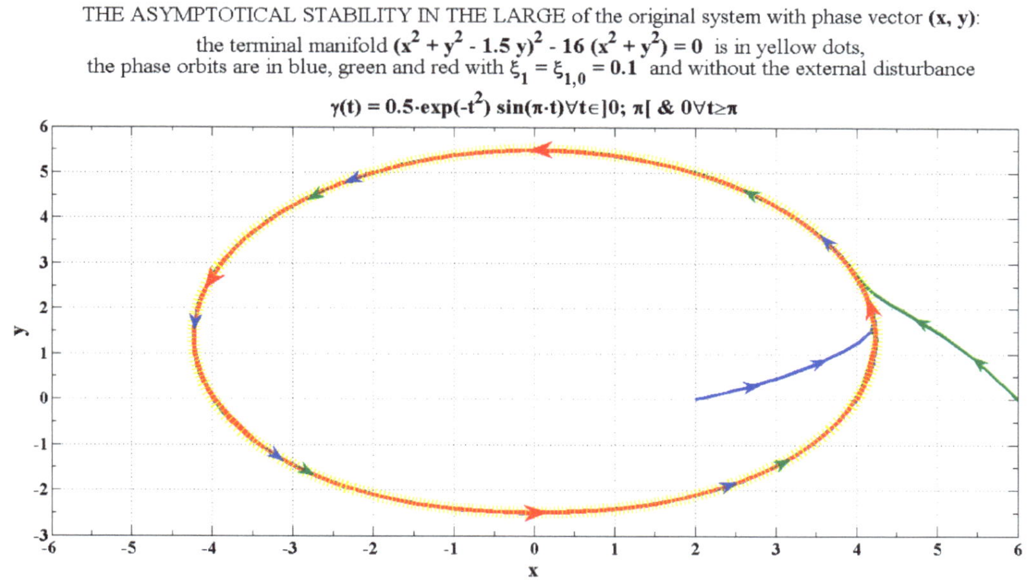
THE ASYMPTOTICAL STABILITY IN THE LARGE of the original system with phase vector (x, y):
the terminal manifold $(x^2 + y^2 - 1.5\,y)^2 - 16\,(x^2 + y^2) = 0$ is in yellow dots,
the phase orbits are in blue, green and red with $\xi_1 = \xi_{1,0} = 0.1$ and without the external disturbance
$\gamma(t) = 0.5 \cdot \exp(-t^2)\,\sin(\pi \cdot t) \forall t \in]0;\ \pi[\ \&\ 0 \forall t \ge \pi$
x
y

GRAPH 19

Substep 6.2 – Asymptotical Stability in the Large and Narrow-Sense Robustness

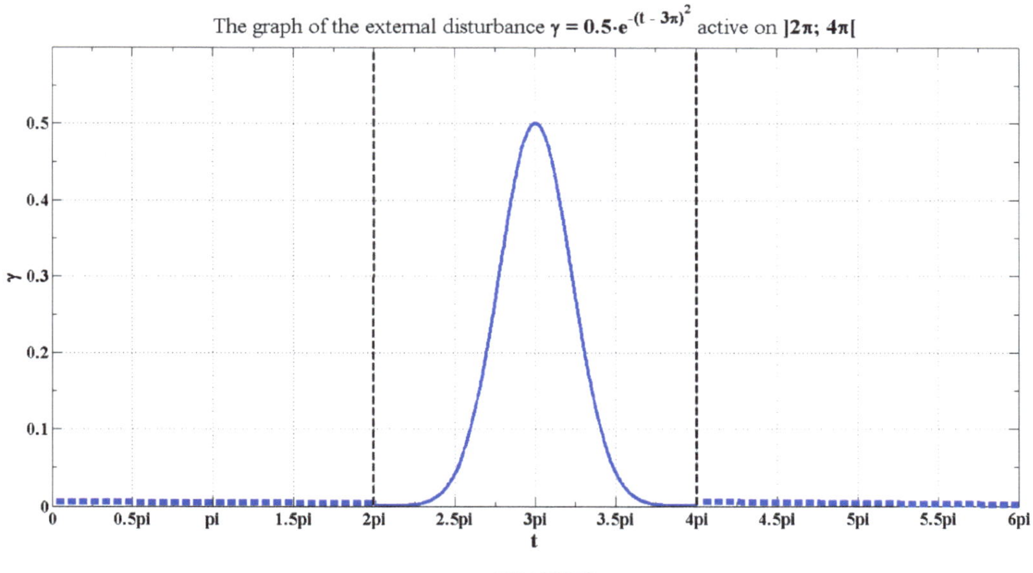
The graph of the external disturbance $\gamma = 0.5 \cdot e^{-(t - 3\pi)^2}$ active on $]2\pi;\ 4\pi[$
t
γ

GRAPH 20

THE ASYMPTOTICAL STABILITY IN THE LARGE AND NARROW-SENSE ROBUSTNESS
of the original system with phase vector **(r, θ)**: the **r** component of integral curves is
in blue, green, red with ξ_1 **= 0.2,** $\xi_{1,0}$ **= 0.1** and the external disturbance

$\gamma(t) = \{0.5\cdot\exp(-(t - 3\pi)^2)\forall t\in]2\pi; 4\pi[\ \&\ 0\forall t\in R^{\geq 0} -]2\pi; 4\pi[\}$

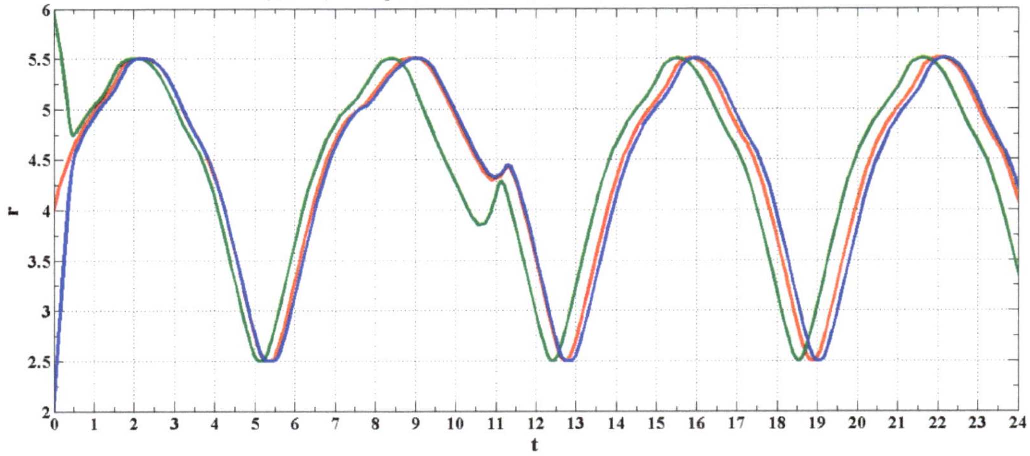

GRAPH 21

THE ASYMPTOTICAL STABILITY IN THE LARGE AND NARROW-SENSE ROBUSTNESS
of the original system with phase vector **(r, θ)**: the **r** component of integral curves is
in blue, green, red with ξ_1 **= 0.05,** $\xi_{1,0}$ **= 0.1** and the external disturbance

$\gamma(t) = \{0.5\cdot\exp(-(t - 3\pi)^2)\forall t\in]2\pi; 4\pi[\ \&\ 0\forall t\in R^{\geq 0} -]2\pi; 4\pi[\}$

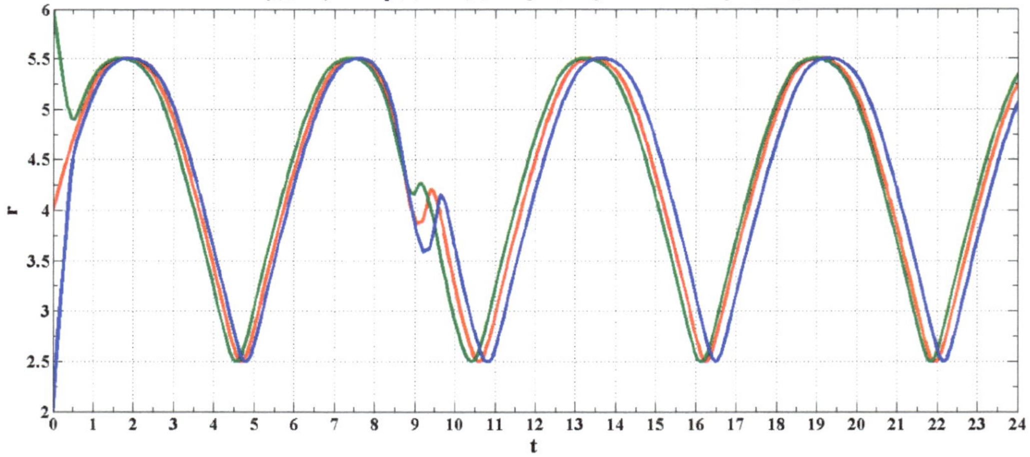

GRAPH 22

THE ASYMPTOTICAL STABILITY IN THE LARGE AND NARROW-SENSE ROBUSTNESS
of the original system with phase vector **(r, θ)**: the terminal manifold **{r = 4 + 1.5· sin(θ)}** is in black, the phase trajectories are in blue, green, red with $\xi_1 = 0.2$, $\xi_{1,0} = 0.1$ and the external disturbance

$\gamma(t) = \{0.5\cdot\exp(-(t - 3\pi)^2) \forall t \in]2\pi; 4\pi[\ \& \ 0 \forall t \in R^{\geq 0} -]2\pi; 4\pi[\}$

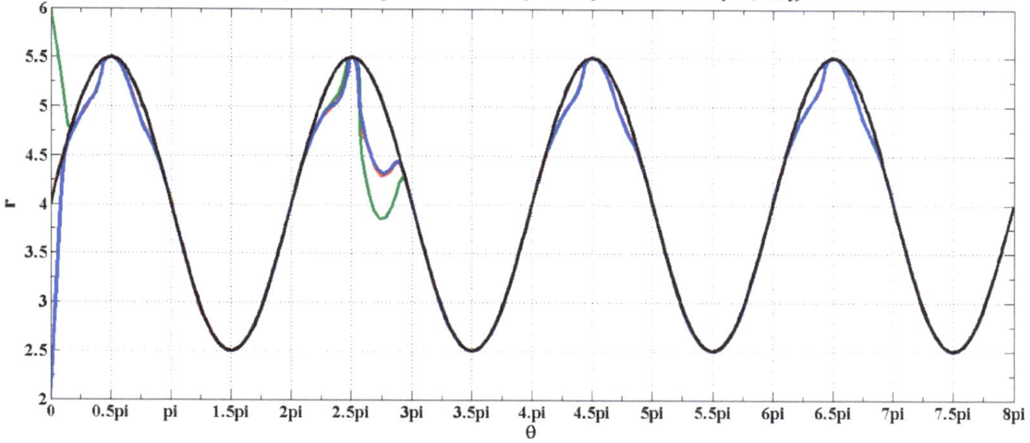

GRAPH 23

THE ASYMPTOTICAL STABILITY IN THE LARGE AND NARROW-SENSE ROBUSTNESS
of the original system with phase vector **(r, θ)**: the terminal manifold **{r = 4 + 1.5· sin(θ)}** is in black, the phase trajectories are in blue, green, red with $\xi_1 = 0.05$, $\xi_{1,0} = 0.1$ and the external disturbance

$\gamma(t) = \{0.5\cdot\exp(-(t - 3\pi)^2) \forall t \in]2\pi; 4\pi[\ \& \ 0 \forall t \in R^{\geq 0} -]2\pi; 4\pi[\}$

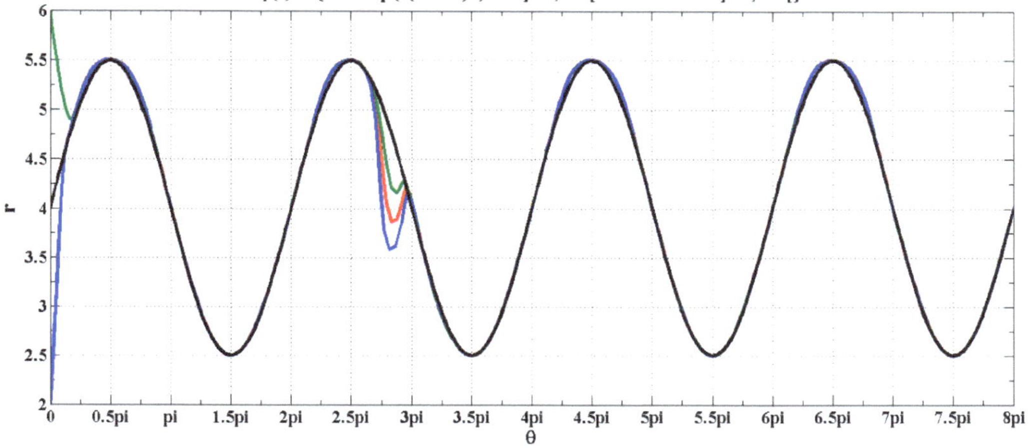

GRAPH 24

THE ASYMPTOTICAL STABILITY IN THE LARGE AND NARROW-SENSE ROBUSTNESS
of the original system with phase vector **(r, θ)**: the θ component of integral curves is in blue, green, red with $\xi_1 = \mathbf{0.2}$, $\xi_{1,0} = \mathbf{0.1}$ and the external disturbance

$$\gamma(t) = \{0.5\cdot\exp(-(t - 3\pi)^2)\forall t\in]2\pi;\ 4\pi[\ \&\ 0\forall t\in R^{\geq 0} -]2\pi;\ 4\pi[\}$$

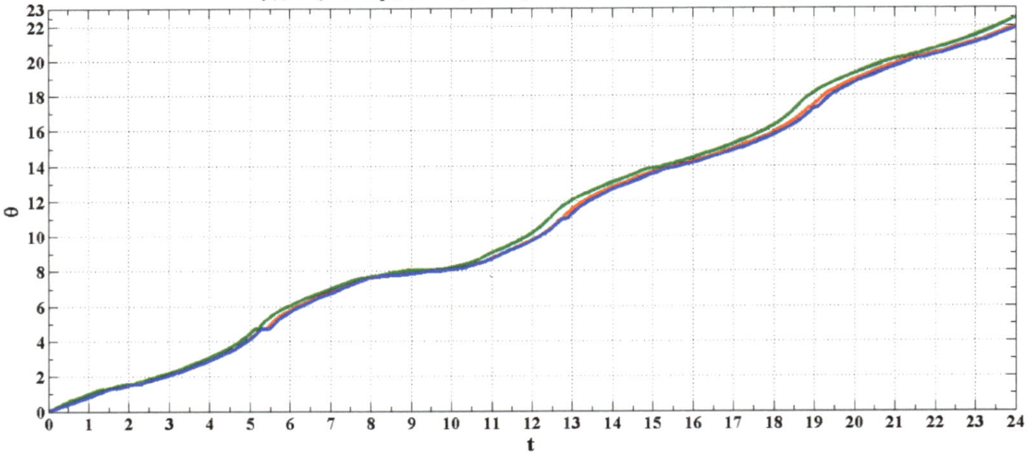

GRAPH 25

THE ASYMPTOTICAL STABILITY IN THE LARGE AND NARROW-SENSE ROBUSTNESS
of the original system with phase vector **(r, θ)**: the θ component of integral curves is in blue, green, red with $\xi_1 = \mathbf{0.05}$, $\xi_{1,0} = \mathbf{0.1}$ and the external disturbance

$$\gamma(t) = \{0.5\cdot\exp(-(t - 3\pi)^2)\forall t\in]2\pi;\ 4\pi[\ \&\ 0\forall t\in R^{\geq 0} -]2\pi;\ 4\pi[\}$$

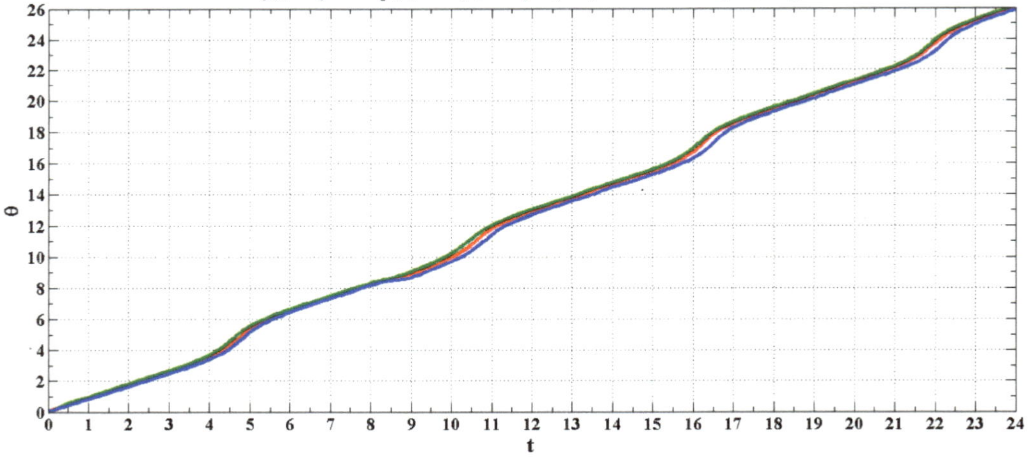

GRAPH 26

THE ASYMPTOTICAL STABILITY IN THE LARGE AND NARROW-SENSE ROBUSTNESS
of the original system with phase vector **(r, θ)**: the control **u** is in blue, green, red with ξ_1 **= 0.2**, $\xi_{1,0}$ **= 0.1** and the external disturbance **γ(t) = {0.5·exp(-(t - 3π)²)∀t∈]2π; 4π[& 0∀t∈R$^{\geq 0}$-]2π; 4π[}**

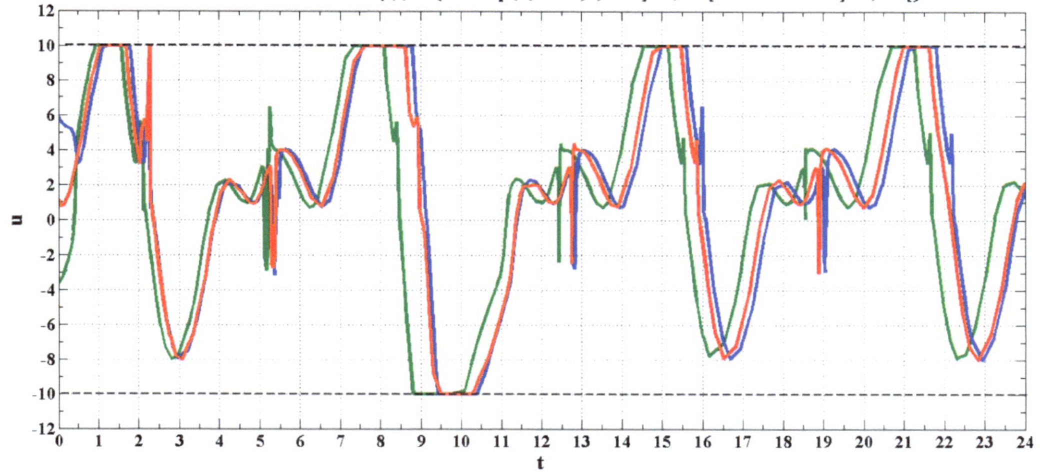

GRAPH 27

THE ASYMPTOTICAL STABILITY IN THE LARGE AND NARROW-SENSE ROBUSTNESS
of the original system with phase vector **(r, θ)**: the control **u** is in blue, green, red with ξ_1 **= 0.05**, $\xi_{1,0}$ **= 0.1** and the external disturbance **γ(t) = {0.5·exp(-(t - 3π)²)∀t∈]2π; 4π[& 0∀t∈R$^{\geq 0}$-]2π; 4π[}**

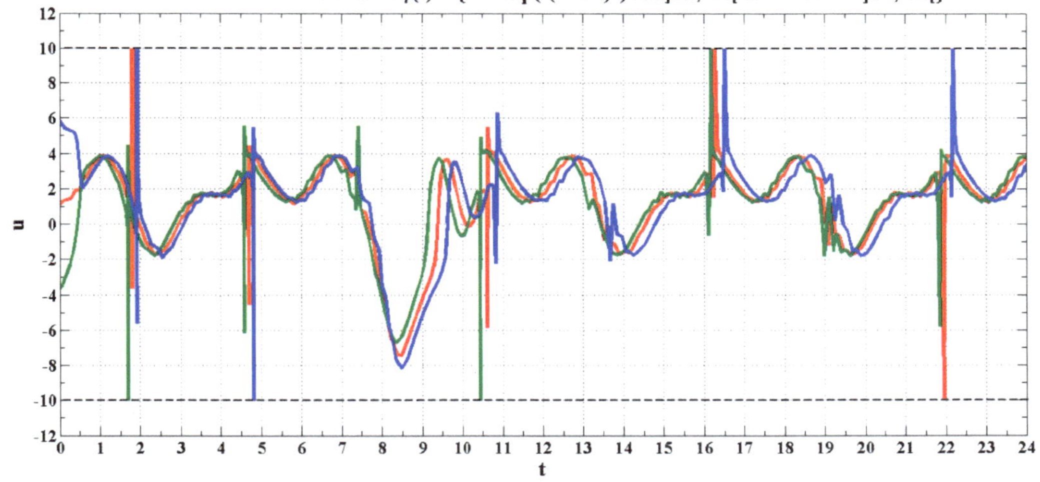

GRAPH 28

THE ASYMPTOTICAL STABILITY IN THE LARGE AND NARROW-SENSE ROBUSTNESS

of the original system with phase vector (**x**, **y**): the terminal manifold $\{(x^2 + y^2 - 1.5\,y)^2 - 16\,(x^2 + y^2) = 0\}$ is in black, the phase orbits are in blue, green, red with $\xi_1 = 0.2$, $\xi_{1,0} = 0.1$ and the external disturbance

$$\gamma(t) = \{0.5 \cdot \exp(-(t - 3\pi)^2) \forall t \in]2\pi;\ 4\pi[\ \&\ 0 \forall t \in R^{\geq 0} -]2\pi;\ 4\pi[\}$$

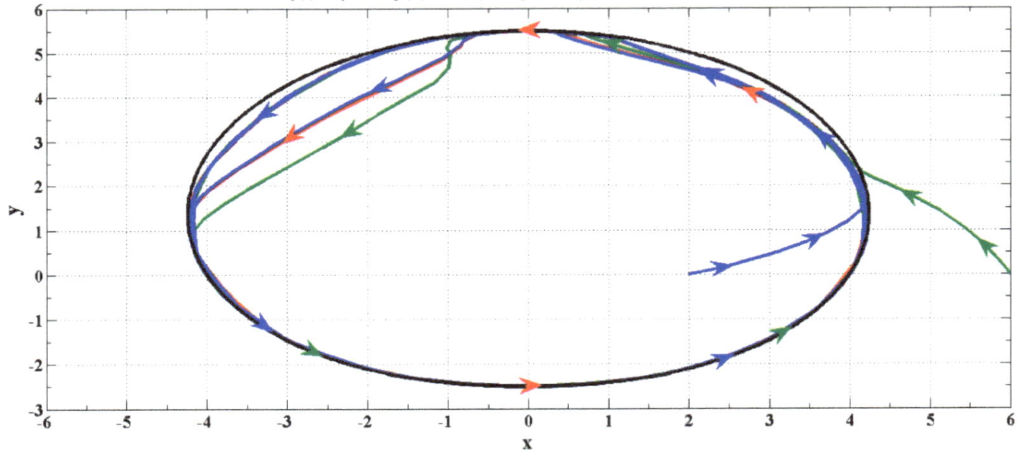

GRAPH 29

THE ASYMPTOTICAL STABILITY IN THE LARGE AND NARROW-SENSE ROBUSTNESS

of the original system with phase vector (**x**, **y**): the terminal manifold $\{(x^2 + y^2 - 1.5\,y)^2 - 16\,(x^2 + y^2) = 0\}$ is in black, the phase orbits are in blue, green, red with $\xi_1 = 0.05$, $\xi_{1,0} = 0.1$ and the external disturbance

$$\gamma(t) = \{0.5 \cdot \exp(-(t - 3\pi)^2) \forall t \in]2\pi;\ 4\pi[\ \&\ 0 \forall t \in R^{\geq 0} -]2\pi;\ 4\pi[\}$$

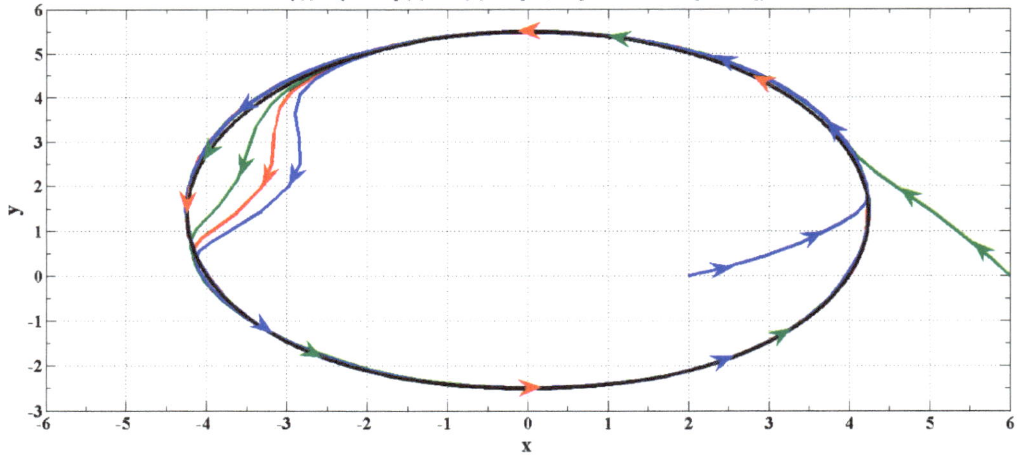

GRAPH 30

THE ASYMPTOTICAL STABILITY IN THE LARGE AND NARROW-SENSE ROBUSTNESS
of the original system with phase vector **(x, y)**: the terminal manifold **$\{(x^2 + y^2 - 1.5\,y)^2 - 16\,(x^2 + y^2) = 0\}$** is in black, the phase orbit **(x(t), y(t))** with **$\xi_1 = 0.2$**, **$\xi_{1,0} = 0.1$** and without the external disturbance is in red

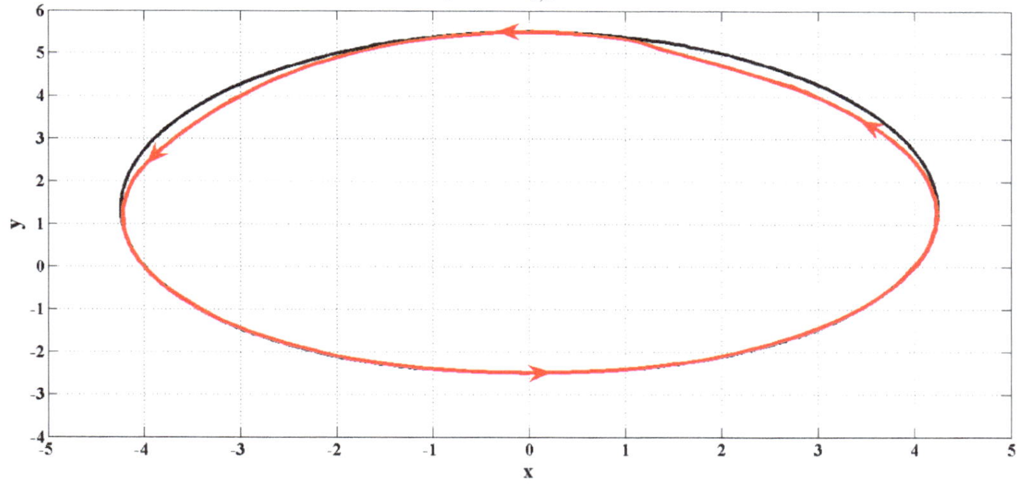

GRAPH 31

THE ASYMPTOTICAL STABILITY IN THE LARGE AND NARROW-SENSE ROBUSTNESS
of the original system with phase vector **(r, θ)**: the function of the quantitative defect of control performance ρ depending on the only argument θ with **$\xi_1 = 0.2$**, **$\xi_{1,0} = 0.1$** and without the external disturbance

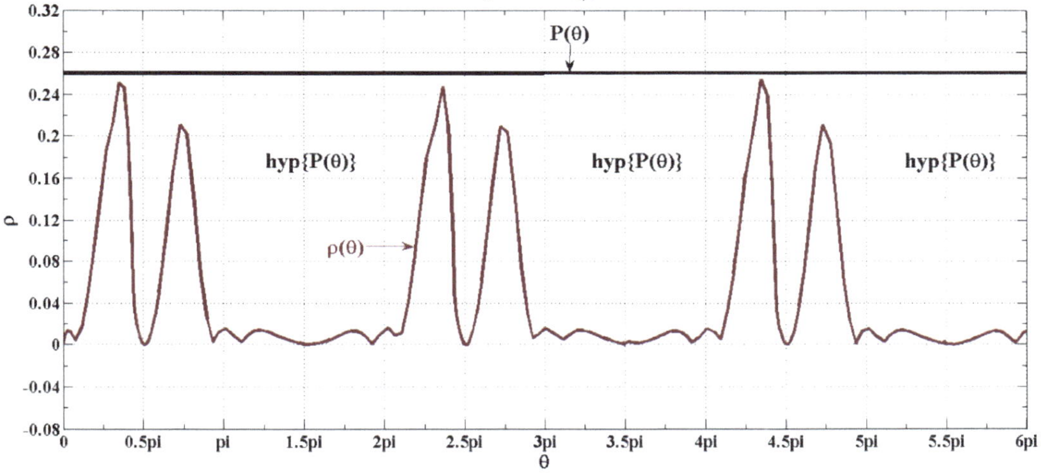

GRAPH 32

THE ASYMPTOTICAL STABILITY IN THE LARGE AND NARROW-SENSE ROBUSTNESS
of the original system with phase vector **(x, y)**: the terminal manifold **$\{(x^2 + y^2 - 1.5\,y)^2 - 16\,(x^2 + y^2) = 0\}$** is in black, the phase orbit **(x(t), y(t))** with ξ_1 **= 0.05**, $\xi_{1,0}$ **= 0.1** and without the external disturbance is in red

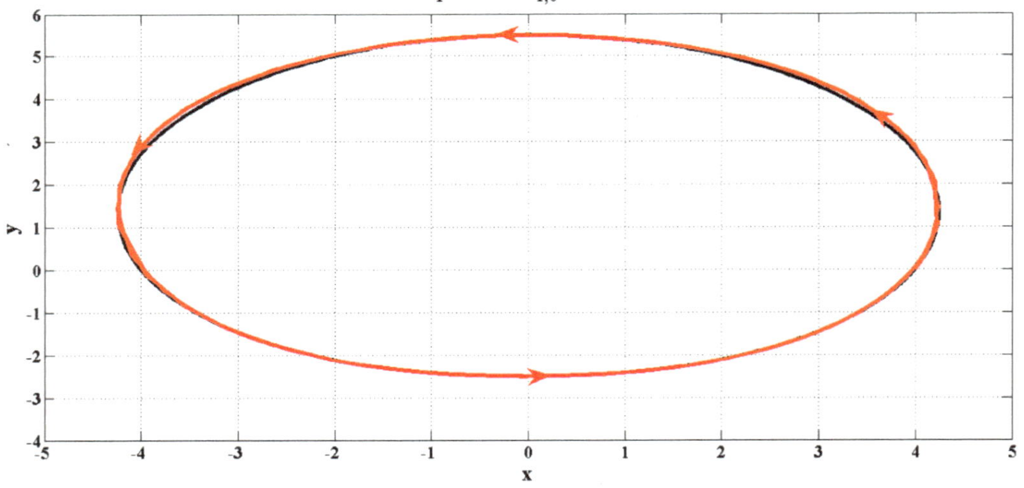

GRAPH 33

THE ASYMPTOTICAL STABILITY IN THE LARGE AND NARROW-SENSE ROBUSTNESS
of the original system with phase vector **(r, θ)**: the function of the quantitative defect of control performance ρ depending on the only argument θ with ξ_1 **= 0.05**, $\xi_{1,0}$ **= 0.1** and without the external disturbance

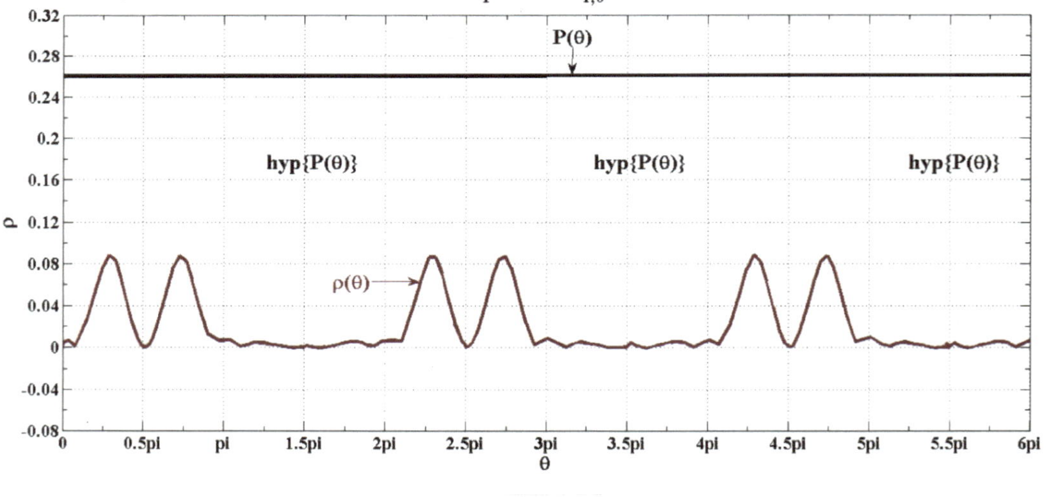

GRAPH 34

Substep 6.3 – Asymptotical Stability in the Large and Wide-Sense Robustness

THE ASYMPTOTICAL STABILITY IN THE LARGE AND WIDE-SENSE ROBUSTNESS
of the original system with phase vector **(r, θ)**: the terminal manifold $\{\mathbf{r = 4 + 1.5 \cdot sin(\theta)}\}$ is in black,
the components **(r(t), θ(t))** of integral curves are in blue, green, red with $\xi_1 = \mathbf{0.2}$, $\xi_{1,0} = \mathbf{0.1}$, the external disturbance
$\gamma(t) = \{0.5 \cdot \exp(-(t - 3\pi)^2) \forall t \in]2\pi; 4\pi[\ \& \ 0 \forall t \in R^{\geq 0} -]2\pi; 4\pi[\}$ and the partially lost control resources $\mathbf{u} \in [-7; 7]$

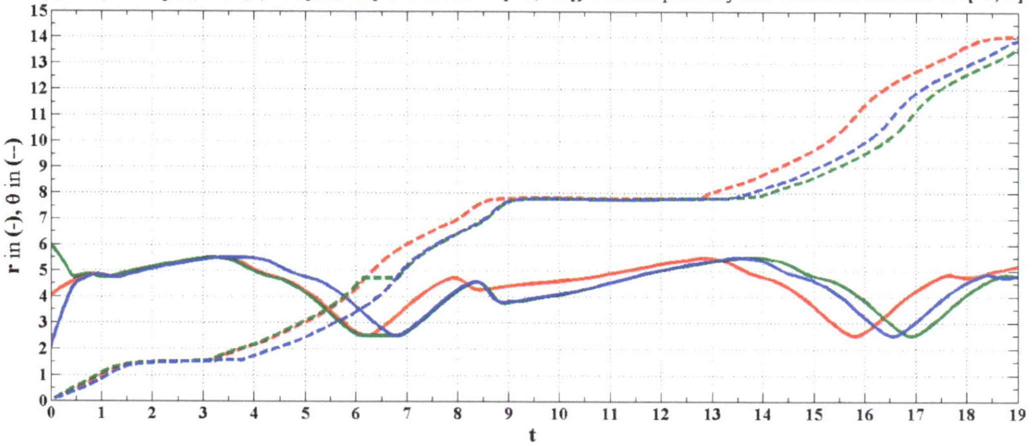

GRAPH 35

THE ASYMPTOTICAL STABILITY IN THE LARGE AND WIDE-SENSE ROBUSTNESS
of the original system with phase vector **(r, θ)**: the terminal manifold $\{\mathbf{r = 4 + 1.5 \cdot sin(\theta)}\}$ is in black,
the components **(r(t), θ(t))** of integral curves are in blue, green, red with $\xi_1 = \mathbf{0.05}$, $\xi_{1,0} = \mathbf{0.1}$, the external disturbance
$\gamma(t) = \{0.5 \cdot \exp(-(t - 3\pi)^2) \forall t \in]2\pi; 4\pi[\ \& \ 0 \forall t \in R^{\geq 0} -]2\pi; 4\pi[\}$ and the partially lost control resources $\mathbf{u} \in [-2; 2]$

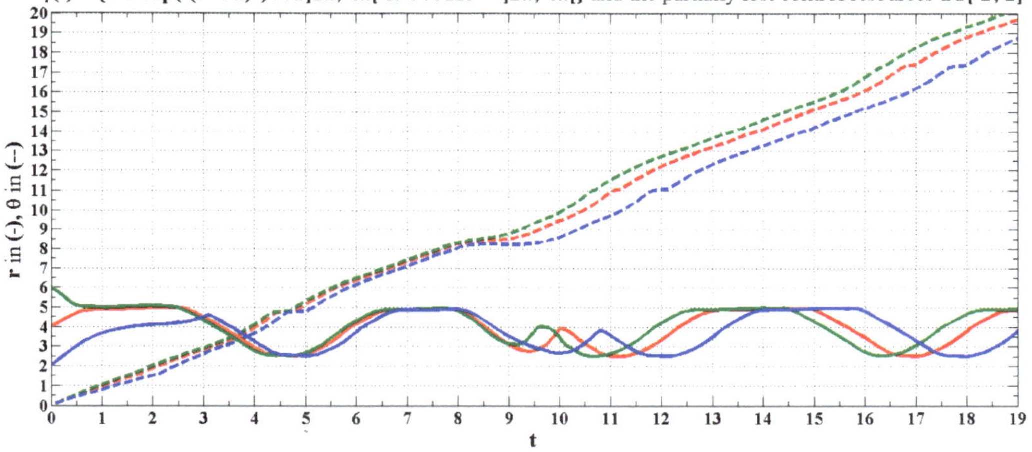

GRAPH 36

THE ASYMPTOTICAL STABILITY IN THE LARGE AND WIDE-SENSE ROBUSTNESS
of the original system with phase vector **(r, θ)**: the terminal manifold **{r = 4 + 1.5· sin(θ)}** is in black,
the phase orbits are in blue, green, red with $\xi_1 = 0.2$, $\xi_{1,0} = 0.1$, the external disturbance

$\gamma(t) = \{0.5\cdot\exp(-(t - 3\pi)^2)\forall t\in]2\pi; 4\pi[\ \&\ 0\forall t\in R^{\geq 0} -]2\pi; 4\pi[\}$ and the partially lost control resources $u\in[-7; 7]$

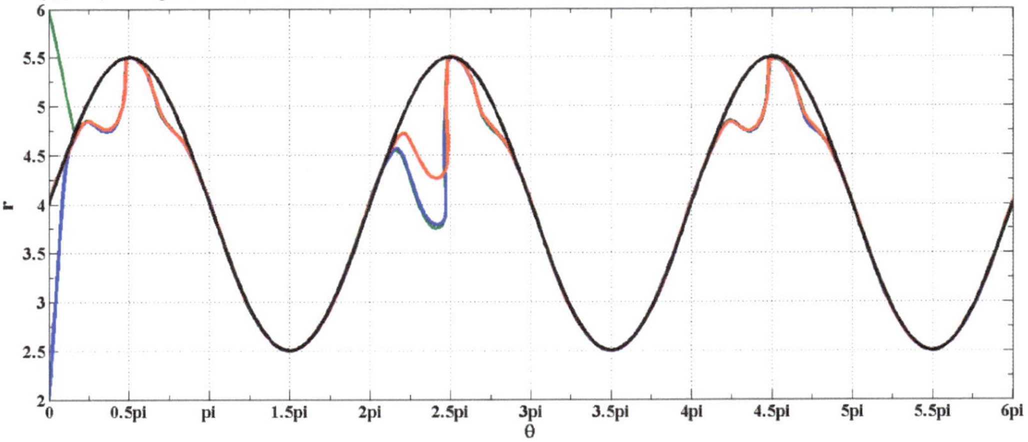

GRAPH 37

THE ASYMPTOTICAL STABILITY IN THE LARGE AND WIDE-SENSE ROBUSTNESS
of the original system with phase vector **(r, θ)**: the terminal manifold **{r = 4 + 1.5· sin(θ)}** is in black,
the phase orbits are in blue, green, red with $\xi_1 = 0.05$, $\xi_{1,0} = 0.1$, the external disturbance

$\gamma(t) = \{0.5\cdot\exp(-(t - 3\pi)^2)\forall t\in]2\pi; 4\pi[\ \&\ 0\forall t\in R^{\geq 0} -]2\pi; 4\pi[\}$ and the partially lost control resources $u\in[-2; 2]$

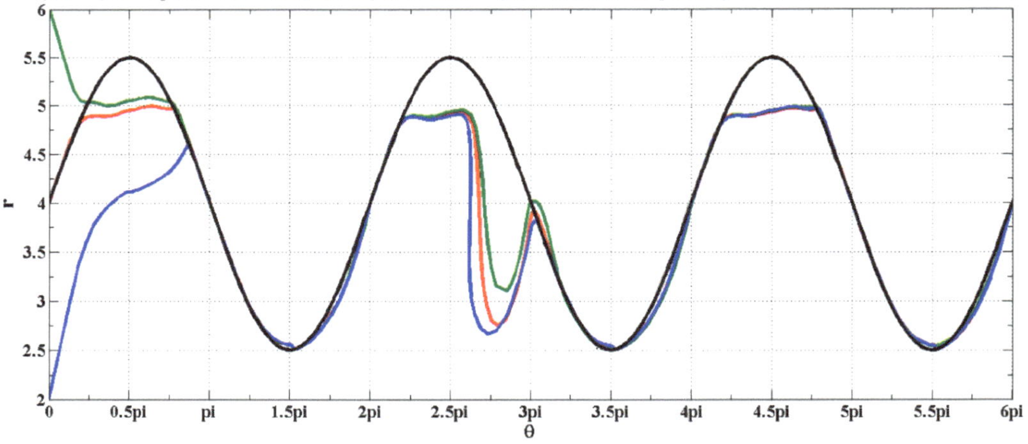

GRAPH 38

THE ASYMPTOTICAL STABILITY IN THE LARGE AND WIDE-SENSE ROBUSTNESS of the original system with phase vector **(r, θ)**: the terminal manifold **{r = 4 + 1.5· sin(θ)}** is in black, the control **u(θ)** is in blue, green, red with ξ_1 **= 0.2**, $\xi_{1,0}$ **= 0.1**, the external disturbance

$\gamma(t) = \{0.5 \cdot \exp(-(t - 3\pi)^2) \forall t \in]2\pi; 4\pi[$ & $0 \forall t \in R^{\geq 0} -]2\pi; 4\pi[\}$ and the partially lost control resources **u**$\in$**[-7; 7]**

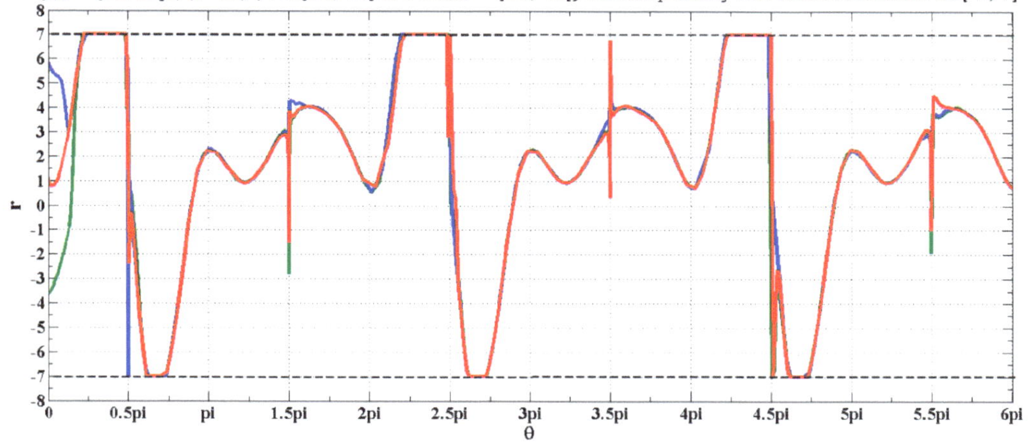

GRAPH 39

THE ASYMPTOTICAL STABILITY IN THE LARGE AND WIDE-SENSE ROBUSTNESS of the original system with phase vector **(r, θ)**: the terminal manifold **{r = 4 + 1.5· sin(θ)}** is in black, the control **u(θ)** is in blue, green, red with ξ_1 **= 0.05**, $\xi_{1,0}$ **= 0.1**, the external disturbance

$\gamma(t) = \{0.5 \cdot \exp(-(t - 3\pi)^2) \forall t \in]2\pi; 4\pi[$ & $0 \forall t \in R^{\geq 0} -]2\pi; 4\pi[\}$ and the partially lost control resources **u**$\in$**[-2; 2]**

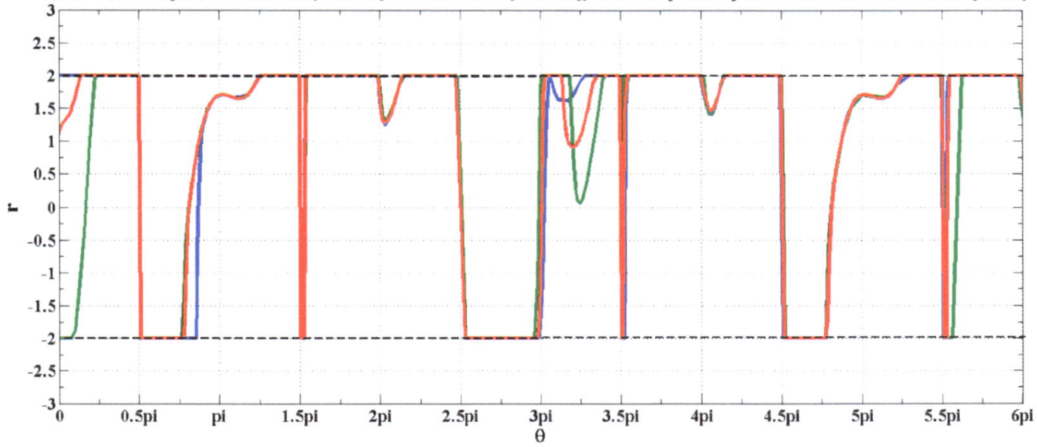

GRAPH 40

THE ASYMPTOTICAL STABILITY IN THE LARGE AND WIDE-SENSE ROBUSTNESS

of the original system with phase vector **(x, y)**: the terminal manifold **$\{(x^2 + y^2 - 1.5\ y)^2 - 16\ (x^2 + y^2) = 0\}$** is in black, the phase orbits are in blue, green, red with **$\xi_1 = 0.2$, $\xi_{1,0} = 0.1$**, the external disturbance

$\gamma(t) = \{0.5 \cdot \exp(-(t - 3\pi)^2) \forall t \in]2\pi; 4\pi[\ \&\ 0 \forall t \in R^{\geq 0} -]2\pi; 4\pi[\}$ and the partially lost control resources **$u \in [-7; 7]$**

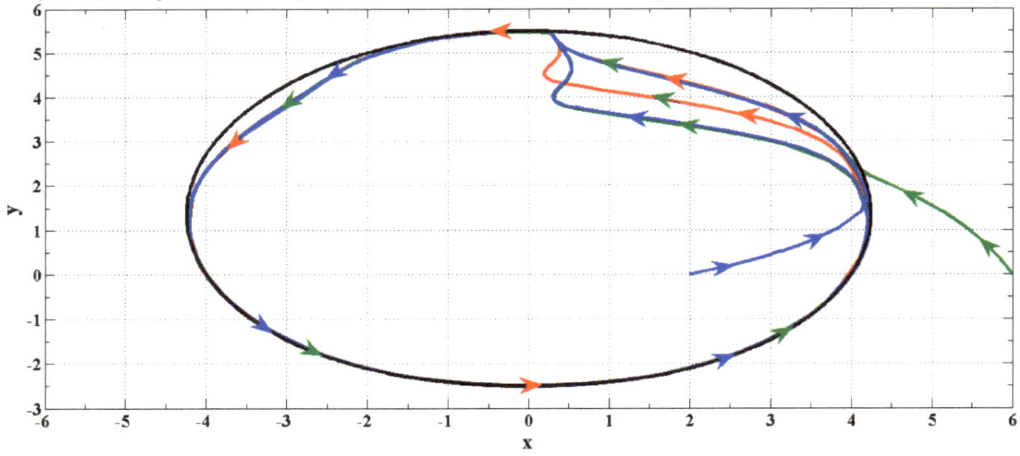

GRAPH 41

THE ASYMPTOTICAL STABILITY IN THE LARGE AND WIDE-SENSE ROBUSTNESS

of the original system with phase vector **(x, y)**: the terminal manifold **$\{(x^2 + y^2 - 1.5\ y)^2 - 16\ (x^2 + y^2) = 0\}$** is in black, the phase orbits are in blue, green, red with **$\xi_1 = 0.05$, $\xi_{1,0} = 0.1$**, the external disturbance

$\gamma(t) = \{0.5 \cdot \exp(-(t - 3\pi)^2) \forall t \in]2\pi; 4\pi[\ \&\ 0 \forall t \in R^{\geq 0} -]2\pi; 4\pi[\}$ and the partially lost control resources **$u \in [-2; 2]$**

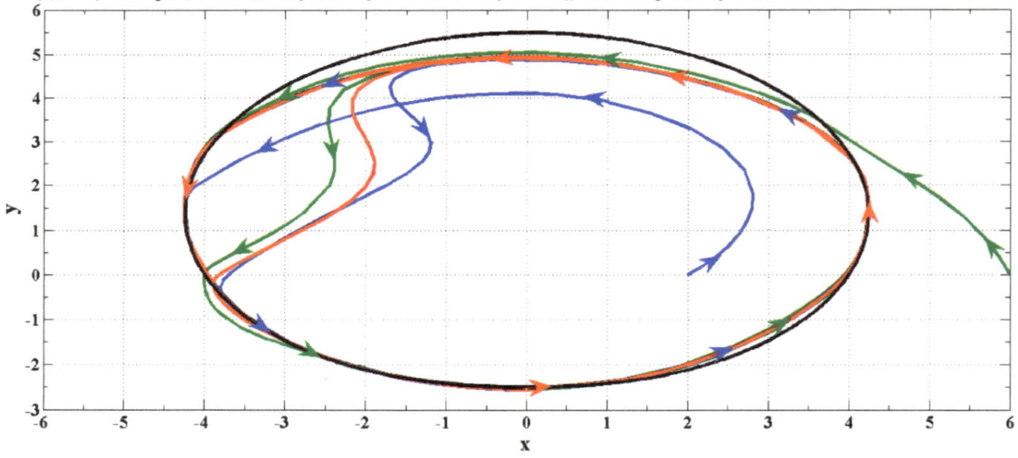

GRAPH 42

THE ASYMPTOTICAL STABILITY IN THE LARGE AND WIDE-SENSE ROBUSTNESS
of the original system with phase vector **(x, y)**: the terminal manifold $\{(x^2 + y^2 - 1.5\,y)^2 - 16\,(x^2 + y^2) = 0\}$ is in black, the phase orbit **(x(t), y(t))** with $\xi_1 = 0.2$, $\xi_{1,0} = 0.1$, the partially lost control resources $u \in [-7; 7]$ and without the external disturbance is in red

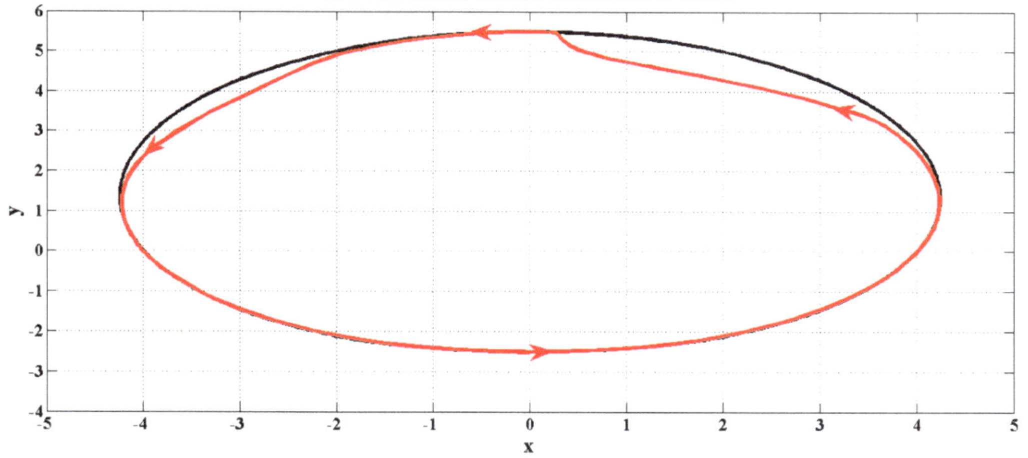

GRAPH 43

THE ASYMPTOTICAL STABILITY IN THE LARGE AND WIDE-SENSE ROBUSTNESS
of the original system with phase vector **(r, θ)**: the function of the quantitative defect of control performance ρ depending on the only argument θ with $\xi_1 = 0.2$, $\xi_{1,0} = 0.1$, the partially lost control resources $u \in [-7; 7]$ and without the external disturbance

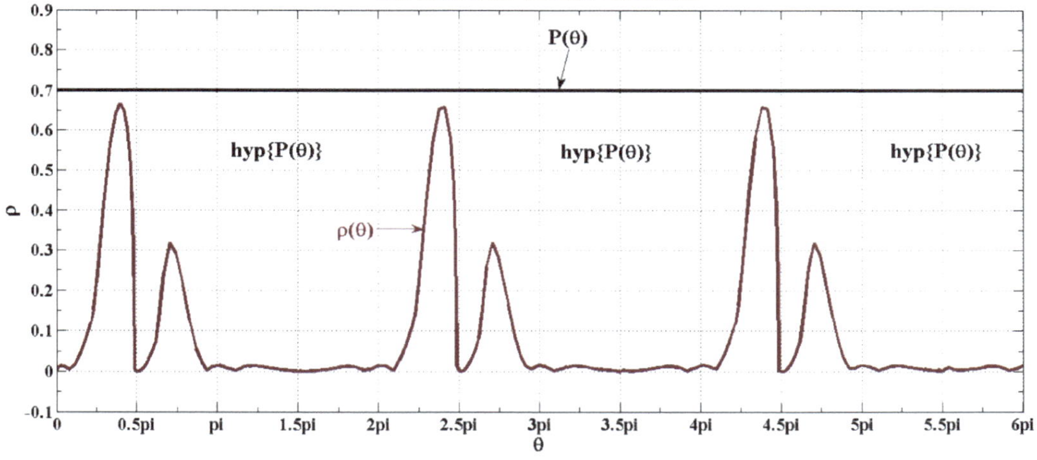

GRAPH 44

THE ASYMPTOTICAL STABILITY IN THE LARGE AND WIDE-SENSE ROBUSTNESS
of the original system with phase vector **(x, y)**: the terminal manifold **$\{(x^2 + y^2 - 1.5\,y)^2 - 16\,(x^2 + y^2) = 0\}$** is in black, the phase orbit **(x(t), y(t))** with ξ_1 **= 0.05**, $\xi_{1,0}$ **= 0.1**, the partially lost control resources **u**$\in$**[-2; 2]** and without the external disturbance is in red

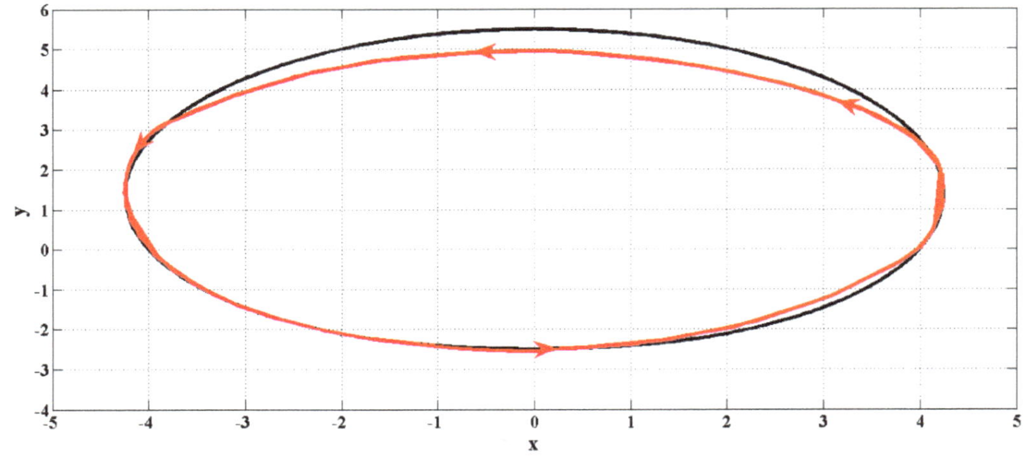

GRAPH 45

THE ASYMPTOTICAL STABILITY IN THE LARGE AND WIDE-SENSE ROBUSTNESS
of the original system with phase vector **(r, θ)**: the function of the quantitative defect of control performance ρ depending on the only argument θ with ξ_1 **= 0.05**, $\xi_{1,0}$ **= 0.1**, the partially lost control resources **u**$\in$**[-2; 2]** and without the external disturbance

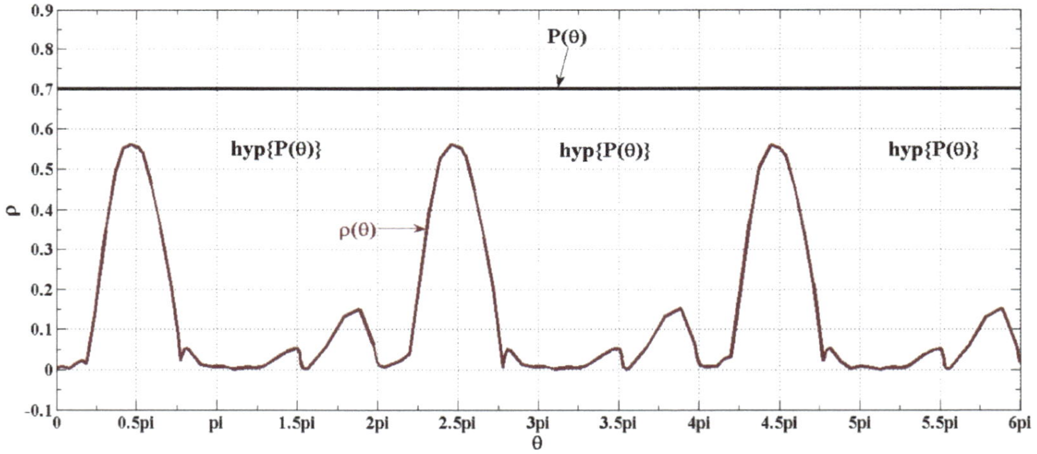

GRAPH 46

It is very useful to peruse the graphs carring the data of the computer simulations collating them.

STEP 7 - MAKING THE CONTROL ADAPTIVE

The goal we have been pursuing on the first six steps is to synthesize the control law making the original dynamic system, first, achieve the control aim and, second, become asymptotically stable in the large and wide-sense robust. The original system with the obtained control law is the base, which the adaptive control will be constructed on. This approach ensures the double-layer protection of the system. If the advanced second layer, namely the adaptive control, fails, then the intrinsic basic layer, the asymptotic stability in the large combined with the wide-sense robustness, bears the brunt maintaining the system survivability.

We understand the adaptation in the sense of the ability of the control to adapt to the unpredictable changes of the parameters of the terminal manifold, the plant and external perturbations by satisfying some adaptation law in the form of an equality/inequality or maintaining the extreme of some functional related to control quality. So we deal with Model Identification Adaptive Control (MIAC) [15].

FIGURE 5. Model Identification Adaptive Control with Search for a Forced van der Pol System

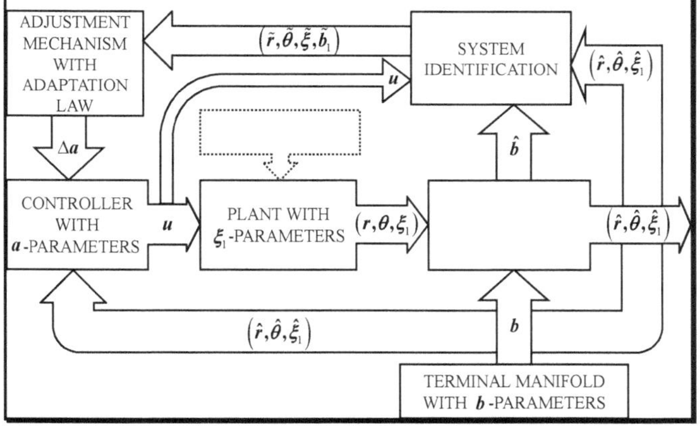

Here $(\hat{r},\hat{\theta},\hat{\xi}_1,\hat{b})$ and $(\tilde{r},\tilde{\theta},\tilde{\xi},\tilde{b}_1)$ are the estimates and the identified values of (r,θ,ξ_1,b) output by the sensors and the system identification respectively; Δa is some correction to a^0 produced by the adjustment mechanism. Let the adaptation law be the requirement to satisfy the constraint on the control $u \in [u_{\min}; u_{\max}]$ at any admissible variations of the parameters $b=(b_1,b_2)\in B$, $\xi_1 \in \Xi$. According to (3) this adaptation law can be expressed as

$$\exists a = a^0 + \Delta a \in A : u_{\min} \le u'(\tilde{r},\tilde{\theta},\tilde{\xi},a,\tilde{b}_1) \le u_{\max}. \quad (134)$$

6. CONCLUSION

We have achieved the following objectives listed below in a particularized form

1) to make the geometric-topological representation of nonlinear non-autonomous parametric differential inclusions in the framework of fiber bundles and foliations;

2) to develop the general procedure of the utilization of Lyapunov functions for the parametric differential inclusions through the canonizing diffeomorphisms defined by the full sets of their parametric first integrals;

3) to investigate

a) the asymptotic stability in the sense of Lyapunov of particular solutions of free dynamic systems as an example of local stability and

b) the global asymptotic stability of the parametric differential inclusions and free dynamic systems as their restrictions;

4) to demonstrate the power and universality of the Poincare's approach to the Lyapunov's second method in control theory having designed the wide-sense robust, adaptive terminal control law.

Let us pay more detailed attention to the points **1)** and **2)**.

First, the geometric-topological structure of the parametric differential inclusions is hierarchical. Each point $\hat{\xi}$ of the manifold Ξ of first-grade or absolutely independent vector of parameters ξ generates a free dynamic system within some given inclusion. The extended phase space or the motion space of the system contains (n) one-codimensional smooth foliations, the leaves of which intersect each other transversely. These leaves are the elementary "bricks" of the space. The intersection of (n) leaves of different foliations gives us an integral curve. The system has one more absolutely independent vector of parameters, namely $c=(c_1,\dots,c_n)\in R^n$ that is the right-hand sides of the vector equation (12) for the full set of first integrals. Generally it belongs to some n-dimensional manifold $C\subseteq R^n$. It has also the second-grade independent vector of parameter x_0 being the initial points of integral curves and depending on c. The vector of parameters c creates the quotient space $X_{t_0}/L_{x_i}\left(\hat{c}_i,\hat{\xi}\right)$ from the manifold X_{t_0} of the initial points of the phase vector x, where $L_{x_i}\left(\hat{c}_i,\hat{\xi}\right)$ are $n-1$-dimensional manifolds, formed by the intersection of the leaves and the hyperplane $\{t=t_0\}$. The one-codimensional foliations "stem from" the elements $L_{x_i}\left(\hat{c}_i,\hat{\xi}\right)$ of the quotient space $X_{t_0}/L_{x_i}\left(\hat{c}_i,\hat{\xi}\right)$ of the manifold X_{t_0}. The structure of fiber bundles can be easily constructed from the structure of foliations if we consider the manifolds of the vectors of parameters Ξ and R^1 or $X_{t_0}/L_{x_i}\left(\hat{c}_i,\hat{\xi}\right)$ as the base spaces.

Second, after we have found the canonical form of the representation of the inclusions it has not been difficult to see the obvious fact formulated as the Theorem 1. Unfortunately, it confirms our worst apprehensions: the Lyapunov functions are some kind of artificial creatures having no direct bearings on the intrinsic properties of the concrete parametric differential inclusions expressed through their parametric first integrals and right-hand sides. Now it is clear why constructing or finding Lyapunov functions for some concrete nonlinear non-autonomous parametric differential inclusion of general form is incredibly difficult problem: *neither right-hand sides nor even first integrals give us any hint at how to address the problem*. Moreover, some specific Lyapunov function can be successfully applied with the same successful results to the absolutely different differential inclusions. And on the contrary, a given differential inclusion can have at least an infinite countable set of Lyapunov functions. They can argue that this ascertainment is valid only for the canonical forms of the parametric differential inclusions. But if we try to go back to the variables x using the inverse canonizing diffeomorphism, which is generated by the vector equation (12) for the parametric first integrals of the initial inclusion (1), then it is possible to receive the Lyapunov functions that will take account of the ones. Even if we are of success on this way, although there are serious doubts about that, it is very difficult to ignore the very strong fact presented by the Theorem 1. W*e can do completely without Lyapunov functions, which have proved to be the superfluous tool in the canonical case, to determine the asymptotical stability not even of a particular integral curve of some particular system of differential equations but all the integral curves of the whole parametric differential inclusion (1), within of which the above-mentioned system corresponds just to a single point of* m*-dimensional Euclidian space*.

The third point of the objectives that we would like to discuss actually is the prospects of the Lyapunov's second method for the investigation of stability in the light of our discovery. Does it downplay the significance of the method of Lyapunov functions? The answer is: by no means. First, because its theory is well developed and can be successfully used in the proofs of theorems and other theoretical reasoning including at control designing. This is in fact just what we have demonstrated in this research. Second, the downsides of the method have been known for a long time. What was not known this was the reason why? Now we are aware of it and this fact entails the understanding that the Lyapunov's second method provides us only with sufficient conditions of asymptotic stability. This means there is a good chance to encounter the following unpleasant phenomenon: directly applying the main theorem of the Lyapunov's second method to a given parametric differential inclusion in its canonical form, which is actually asymptotically stable, does not evince any satisfactory results on the existence of this property in the inclusion. The typical case of such a situation is the presence of mosaic attractors, introduced in [5], in the motion spaces of dynamic systems. The structure of their mosaic patterns can be very complex. The Lyapunov functions are quite insensitive to asymptotic stability created by them because they are not able to detect the presence of mosaic attractors in motion spaces. Thus, parametric differential inclusions and their restrictions as dynamic systems or integral curves can be asymptotically stable but it is impossible to apply the Lyapunov second method in such way that the one will confirm this fact. There are two ways out of this situation. The first one is to modify the method in order to add to its "credentials" the feature able to discern mosaic attractors. The second alternative way is to use the Poincaré's approach developed in [5] to investigate the problems of stability.

Now turn our attention to the fourth point of the objectives dealing with the design of adaptive terminal control law delivering the asymptotical stability in the large and wide-sense robustness to the original dynamic system. As we have said in the beginning of Step 7 the two above-mentioned properties create the intrinsic basic layer of the protection of the system, on which the second advanced layer, the adaptivity, is based. All man-made physical systems are supposed to accomplish some predesignated missions. Furthermore, they must "demonstrate certain persistency and stubbornness" in their goal-oriented functioning despite all unfavorable factors.

In other words, these systems should be designed to keep themselves on the preplanned tracks that lead to the goals of the missions

1) fighting back and compensating the measurable, identifiable external disturbances and adverse changes of environmental conditions respectively;
2) staying almost insensible to immeasurable, unidentifiable disturbances and changes.

The first feature relates to adaptation, the second one pertains to stability and robustness. In this book the main accent is put on the latter. The Definition 10 has introduced two important concepts: the qualitative (topological) equivalency and the quantitative defect of control performance. Both of them are directly linked to the problems of attainability and controllability. For dynamic systems with control narrow- and wide-sense robustness means that attainability and controllability are preserved topologically to the full extent and quantitatively to some admissible degree of error.

Our discussion would not be complete if we would not say a couple of words about wide-sense robustness. The partial or full loss of control resources is a major reason of the catastrophic consequences of all kinds of failures and malfunctions. To put it bluntly, this is one of the worst scenarios - if not the worst one that can happen to the system on its way to the control goal. Endowing the systems with the intrinsic property of wide-sense robustness should be on the priority list of operational performance requirements.

REFERENCES

1. Sparavalo, M. K.: The Lyapunov Concept of Stability from the Standpoint of Poincaré Approach: General Procedure of Utilization of Lyapunov Functions for Non-Linear Non-Autonomous Parametric Differential Inclusions. arXiv.org > cs > arXiv:1403.5761, 2014.
2. Takahara, M.: Geometry, Topology and Physics. Taylor & Francis, 2nd Edition, 2003.
3. MacLane, S., Moerdijk, I.: Sheaves in Geometry and Logic. Springer, 1994.
4. Berezin, F.: Introduction in Superanalysis. Springer, 1987.
5. Sparavalo, M. K.: The Topological Analysis of Non-specific Motions and Its Applications to Control Problems, D.Sc. Thesis in Physics and Mathematics, Glushkov Institute of Cybernetics of National Academy of Sciences of Ukraine, Kiev, 1995. (in Russian)
6. Sparavalo, M. K.: The Geometry of Vector-Fields Spaces without Critical Elements: Differential-Topological Approach to the Analysis Problem in Control Theory. *Problemy Upravleniya i Informatiki, no. 1*, 1995, pp. 1-25. (In Russian)
7. Demidovich, B. P.: Lectures on Mathematical Theory of Stability, Nauka, Moscow, 1967. (in Russian)
8. Sparavalo, M. K.: Robust Adaptive Control for Circadian Dynamics: Poincaré Approach to Backstepping Method. arXiv.org > cs > arXiv:1310.2306, 2013.
9. Sparavalo, M. K.: A Method of Goal-Oriented Formation of Local Topological Structure of Co-Dimension One Foliations for Dynamic Systems with Control. J. of Automation & Information Sci., 25(2), 1992, 65-71.
10. Sparavalo, M. K.: Attractors of Dynamical Systems with Control: Topology of Purposeful Formation. Ukrainian Mathematical Journal January 1995, Vol. 7, Issue 1, 152-156.
11. Kokotović, P. V: The Joy of Feedback: Nonlinear and Adaptive. *Control Systems Magazine, IEEE* **12** (3), 1992, 7–17.
12. Kristić, M., Kanellakopoulos, I., Kokotović: P. Nonlinear and Adaptive Control Design. John Wiley & Sons, Inc., 1995.
13. Rumyantsev, V. V. On the Stability of Motion with Respect to Part of the Variables, *Vestnik Moskov. Univ. Ser. Mat. Mekh. Fiz. Astron. Khim.,* 4, 9-16.
14. Vorotnikov V. I.: Partial Stability and Control. Birkhäuser, 1998.
15. Cheng, B. H. C. et al. (Eds.): Software Engineering for Self-Adaptive Systems, LNCS 5525, pp.48-70, 2009.

www.ingramcontent.com/pod-product-compliance
Lightning Source LLC
Chambersburg PA
CBHW041730210326
41598CB00008B/830
* 9 7 8 0 6 9 2 6 9 4 2 4 4 *